Samuel Webber

Manual of power for machines, shafts and belts

with the history of cotton manufacture in the United States

Samuel Webber

Manual of power for machines, shafts and belts
with the history of cotton manufacture in the United States

ISBN/EAN: 9783744799263

Printed in Europe, USA, Canada, Australia, Japan

Cover: Foto ©Suzi / pixelio.de

More available books at **www.hansebooks.com**

LIBRARY

2.

MANUAL OF POWER

FOR

MACHINES, SHAFTS, AND BELTS.

WITH THE

HISTORY OF COTTON MANUFACTURE

IN THE

UNITED STATES.

BY

SAMUEL WEBBER, C. E.

NEW YORK:
D. APPLETON AND COMPANY,
1, 3, AND 5 BOND STREET.
1891.

COPYRIGHT BY
SAMUEL WEBBER.
1879.

TO

HON. E. A. STRAW,

PRESIDENT OF THE NEW ENGLAND COTTON MANUFACTURERS' ASSOCIATION,

THIS COLLECTION OF TESTS, COMMENCED AT HIS REQUEST,

IS RESPECTFULLY DEDICATED,

BY THE COMPILER.

PREFACE TO REVISED EDITION.

In preparing for the press a second and enlarged edition of my "Tests of Power," a somewhat more extended preface seems necessary, in order to explain what have appeared to many persons to be discrepancies in the first edition, and also to give a short explanation of the manner in which the results were attained.

The dynamometer used in the great majority of the tests was designed by Hon. E. A. Straw, of Manchester, New Hampshire, on the same principle as the one originally invented by Samuel Batchelder, Esq., at Saco, Maine, in the year 1836, but contained the modifications of the hydraulic regulator and second transmitting shaft, which were added to the original machine by James B. Francis, C. E., of Lowell. The principle of the machine may be described as follows: A shaft containing the receiving pulley passes through the fulcrum of a steelyard, having fast on it at the end next the pulley a bevel gear, which forms one side of a "box or compound" gear. On a sleeve at the opposite end of the shaft, and revolving freely around it, are fixed another bevel gear of equal diameter, forming the opposite side of the "box," and a plain gear, which transmits the power to a second shaft parallel with the first, and which carries the delivering pulley, which is thus brought in the same line of transmission as the receiving pulley. Around the steelyard, on either side of the fulcrum, revolve freely two other bevel gears, of equal diameter with those mentioned, which complete the "box" or "compound," and which transmit the motion from the first shaft to the sleeve, from which it is given to the second shaft by a pair of equal gears.

If the dynamometer is put in motion without attaching any machine to the second shaft, the gears revolve around the steelyard without any other resistance than their own friction, and no effect is recorded; but if a belt be carried from the delivering pulley to the pulley on any machine, the resistance caused by such machine tends to act upon the steelyard in such a manner as to give it a motion around its fulcrum.

This steelyard is made of such length that a weight hung at its extremity would describe a circle of 10 feet each revolution, if the steelyard were allowed to rotate on its fulcrum; and consequently the 100 revolutions would move 1,000 feet.

A set of weights are therefore prepared, which are sufficient to hold the steelyard at a level poise when the power is being transmitted through the dynamometer, and each one pound weight is considered to move 1,000 feet in 100 revolutions, or to equal 1,000 pounds moved one foot. The steelyard is also graduated into inches and tenths, and a poise weight prepared, which represents 100 pounds moved one foot in 100 revolutions, for each inch that it is carried out on the steelyard toward its outer end, this poise being 3.84 pounds in actual weight. The weights representing 1,000 pounds are also double the actual weight necessary to represent that sum, as the steelyard would only make 50 revolutions, if left free, while the shaft was making 100.

The dynamometer, being fastened in position and belted properly, is put in motion, the machine to be weighed being driven through it; and the weight necessary to hold the steelyard perfectly level while driving the machine is attached to it in the usual manner. When this balance is properly adjusted, the revolutions of the shafts are counted by a clock driven by a worm and gear attached to one of them, and the apparent weight hung on the steelyard is divided by the number of seconds consumed in making 100 revolutions.

Thus if 11,000 pounds be the marking of the weight hung on the steelyard, and the time of completing 100 revolutions be 10 seconds, the answer to the division is $\frac{11,000}{10} = 1,100$ pounds lifted a foot in one second, or 2-horse power—550 pounds lifted one foot per second being the equivalent of 33,000 pounds lifted one foot per minute, the established standard of a horse power. The amount of weight necessary to balance the friction of the dynamometer itself, when in revolution at the desired speed, is first ascertained and deducted before making such divisions, and will average not far from five per cent. of the whole power consumed, but in all cases in these tests has been actually ascertained and deducted in each instance.

Previous to the commencement of these operations, the dynamometer was compared with the one designed by Mr. Francis for the Locks and Canals Company, of Lowell, with which it was found to agree exactly, and was then further verified by over one thousand tests with a Prony brake, the friction pulley of which was attached to the machine in the place of the ordinary delivering pulley and the arm of the lever loaded with varying weights; while by a series of pulleys the speed of the dynamometer was changed from time to time, so as to vary from 200 to 1,000 revolutions per minute.

During all these tests the steelyard of the dynamometer was found to record accurately the load placed on the brake-lever, plus a certain uniform difference for the friction of the machine, which increased in a regular ratio with the velocity, and agreed very closely with the amount of friction observed by running the dynamometer without a load, and it was therefore decided to adopt the method of deducting the actual observed friction in every test.

The greater part of the tests in the following tables have been taken without any previous preparations of the machines to be weighed, the dynamometer having been attached to them as they were actually in operation in the mills; but there are some few exceptions which should be noted.

All the tests at Manchaug in August, 1871, were of comparatively new machinery, which was in the best possible order, and had been run just about long enough to get fairly eased in its bearings;

and the same remarks will apply to the tests at the Manchester Print Works in June, 1872. The tests at Manville, Rhode Island, were of a new frame in perfect order, kept thoroughly clean and well oiled, but in an unfinished mill, where the atmosphere was damp and cool until the last day of the trials, reference to which will show that the machinery ran with less power on that day; and the same difference will be noticed in the tests at Mount Vernon Mills, Baltimore, as well as the difference caused by the oils used for lubrication in the same trials.

The tests of the Pusey Spindle at Wilmington, Delaware, and those of the Sawyer Spindle at the Appleton Mills, Lowell, were also made under the most favorable circumstances, as were those of the Pearl and Rabbeth Spindles in June, 1873, at the Pacific Mills, and those of the Rabbeth in March, 1873, at the Potomska Mills. The difference due to cleaning and lubrication will be noticed in the tests of spinning September 25-27, 1873, at the Atlantic Mills, and of the throstle frame, A, at the Stark Mills in April and November, 1871.

The difference due to the atmosphere is strikingly shown in the tests of the same spinning frame, August 7th and 8th, 1872, at the Amoskeag Mills.

The differences due to banding may be seen in many places, but in none more strikingly than in the tests at the China Mill in March, 1875; and those caused by tight belts may be seen in the trials at the Ocean Mills, June, 1873.

The matter of banding is one to which it is well worth while to call the attention of spinners, as all tension beyond that actually required to drive the spindle up to its proper speed causes unnecessary friction in the bearings, and wears out both them and the spindles rapidly.

A band should never be tied on so as to be rigid, but should always retain its elasticity; and the same is true in regard to belts; and a little extra attention to these matters is always true economy.

Geared machines, like drawing, speeders, and fly frames, will not of course show these differences; but with spinning every detail must be taken into account to prove a correct conclusion as to the result. It has also been found impossible to test a single loom with absolute correctness, as the dynamometer tended to register the extreme power of the beat, particularly if the speed was high.

It should also be noted that the speed of the mule spindles given is the actual number of revolutions per minute, deducting the time taken up in "running in" the carriage, and not the velocity of the spindle when in operation, and the comparative power of the mules and frames is ascertained by the number of yards of yarn per spindle per minute, as compared with the foot pounds.

This in the mule is equal to the length of stretch, multiplied by the number of stretches per minute, and in the frames by the number of revolutions of the front roll, multiplied by 3.1416 or 3¼, the diameter of the front roll being one inch.

Taken as a whole, however, the results obtained have agreed remarkably with those obtained by the steam-engine indicator, when the summary of the dynamometer results was compared with the indicator cards of the same mill, after allowing for friction of engine, or with the calculation of the effect of the Boyden turbine in cases where water power was used, particularly in the cases shown in the summaries of tests as in mills B and I.

In two cases "Swain" Wheels have been put into mills after getting the required power by the dynamometer, namely, the Whittenton Mills at Taunton, and the Clinton Mills at Woonsocket, and in both cases have agreed exactly with the calculations previously made.

It should also be mentioned that in some of the mills tested, such as those at Rockport, Newburyport, Gloucester, N. J., Haydensville, Mass., and the Stark Mills at Manchester, N. H., much or all of the machinery was quite old, and the tests were made to ascertain the actual consumption of power, with a view to future improvements; while in the tests previously referred to such improvement had commenced, and the trials were for the purpose of demonstrating the prospective saving to be attained.

Such as the records are they are correct, and any apparent discrepancies in them are usually explained at once by a full knowledge of the circumstances.

It should also be remembered that neither very new nor very old machinery can be depended on for a fair average test of the power usually required. In the first case there is a loss of power from extra friction, in the last from extra gyration.

Although the greater part of my tests have been confined to cotton machinery, there are a sufficient number of trials of woolen, worsted, and flax machines to give a fair idea of the usual amount of power required to operate them at the usual velocity, together with a short list of machine tools and miscellaneous machinery. Paper machinery has not been tested, for want of a dynamometer of sufficient capacity.

To the separate tests of machines, I add in this edition the summaries of all the machines, with the power required by them, in a series of mills on different fabrics, and also tables for shafting and belting; the former calculated from the tables of Mr. James B. Francis, the latter from various authorities, as well as from my own experiments, and at the request of several manufacturers add the English tables of twist for yarn, and roving, and for the breaking strength of yarns, together with certain convenient rules for calculating drafts and numbers.

I also add a corrected report of the turbine tests at the Centennial Exposition, which contained a number of small errors in the official publication, though not enough to invalidate the general result.

The work is completed for the present by a historical sketch of the growth and progress of the cotton manufacture in the United States, originally prepared for the American Society of Civil Engineers, as part of a centennial record of various American industries, but the publication of which has been so long delayed, from various causes, that the officers of the Society have consented to its publication in this form.

MANCHESTER, N. H., *January*, 1879.

PREFACE.

These tests, undertaken without any view to publication, may be found deficient in some points; but may be depended on as correct, as far as they go.

They give a representation of the power required by cotton-machinery, as actually in use, under various circumstances, in a large number of mills: some of it new, and in the best running order; some of it very old, and tested with a view to ascertain how much power was lost by using it. Part of the tests have been made to ascertain the variations due to weather, oil, and banding; but the writer believes that the notes will explain sufficiently these different points, and trusts that the information contained, as to the average power actually used, will be of sufficient value to manufacturers to justify their publication.

TABLE OF CONTENTS.

PART I.

	PAGES
EXPLANATORY PREFACE...	3
Power Tests of Cotton Openers and Pickers.............................	12, 56
" " Cards...	16, 59
" " Railway Heads.......................................	18, 59
" " Drawing Frames......................................	20, 59
" " Roving Frames.......................................	22, 60
" " Throstle Spinning...................................	28
" " Ring Spinning, Common Spindle.......................	30, 62
" " " Sawyer "........................	34, 66
" " " Pearl "........................	36, 68
" " " Rabbeth "........................	36, 70
" " " Birkenhead ".......................	38, 70
" " " Excelsior "........................	38, 70
" " " Perry "........................	42, 70
" " " Pusey "........................	42
" " " Miscellaneous Spindles..............	70
" " Mule Spinning.......................................	44, 72
" " Spoolers..	48, 73
" " Twisters..	48, 72
" " Warpers...	48, 73
" " Dressers..	48, 73
" " Looms...	46, 74
" of Miscellaneous Machinery and Tools.......................	49, 78
" of Flax Machinery..	50, 72
" of Wool "	50, 74
" of Worsted "	79
" of Shafting...	52
" of Mills in full..	78
Rules and Tables for Shafting.......................................	88
" " " Belting...	95
Turbine Wheel Tests...	101
Roving and Yarn Tables..	115

PART II.

CHAPTER I.

Commencement of Cotton Manufacture by Machinery—Arkwright—Paul—Hargreaves—Crompton—Wyatt.. 1

CHAPTER II.

First attempts in the United States—Bridgewater—Beverly—Philadelphia—Rhode Island, etc... 7

CHAPTER III.

Samuel Slater—Almy and Brown—Eli Whitney—Pliny Earle—Amos Whittemore.. 15

CHAPTER IV.

1800 to 1812—Rapid Growth—Pawtucket—Paterson—Watertown—New Ipswich, N. H.—Samuel Batchelder—Peterborough, N. H.—Amoskeag Falls—General Statistics—1810... 21

CHAPTER V.

Nathan Appleton—Power Looms—Francis C. Lowell—Patrick T. Jackson—Paul Moody—Waltham—The New England System............................... 28

CHAPTER VI.

1812—Beginning at Fall River—Paterson—Ludlow—North Adams—Matteawan—William Gilmore—Power Looms—Ira Draper—Rotary Temple—Statistics—1820... 34

CHAPTER VII.

1821—Origin of Lowell—Kirk Boott—Nathan Appleton—Paul Moody—Francis C. Lowell—John D. Prince—Samuel L. Dana—Merrimac Manufacturing Company—Hamilton Company—Appleton Company—Samuel Batchelder—First Mill driven by *main belts* by Paul Moody—Nashua—Dover—Chicopee—Tariff of 1824—A. & A. Lawrence—Aza Arnold—Equation Box—Saco—Cohoes—Tariff of 1828—Danforth's Cap-Spindle.. 39

CHAPTER VIII.

Ring-Spinning—John Thorpe—William Mason—Brewster's Speeder—York Manufacturing Company—Samuel Batchelder—Balance Dynamometer—Cotton Crop of 1834—Boston & Lowell Railroad—Patent Office established—Newburyport—Amoskeag Company—Stark Mills—E. B. Bigelow—Counterpane Loom—Statistics—1840.. 47

CHAPTER IX.

Self-Acting Mule—Ira Gay—Pitcher & Brown—William Mason—Richard Roberts—William C. Davo—Smith Mule—Higgins Mule—Potter Mule—Parr & Curtis—Platt Brothers—Wanton Rouse—English Roving Frames—Higgins & Sons—

	PAGE
First Turbine Wheels—Uriah A. Boyden—Lawrence—Essex Company—Atlantic Mills—E. B. Bigelow—Gingham Loom—Increase in size of Mills—Production—1850.	52

CHAPTER X.

Hadley Falls Company—Holyoke—Lewiston—Rapid Growth to 1857—Self-Stripping Card—George Wellman—Horace Woodman—Bag Loom—Cyrus W. Baldwin—Picker Staff—W. W. Dutcher—Railway Evener—D. W. Hayden—George Draper—Pemberton Mills—Census of 1860—Progress of Ten Years............. 58

CHAPTER XI.

The Great Rebellion of 1861—Prostration of Business—Close of the War—Revival of Manufactures, and enormous increase—"*Slasher*" Dresser Introduced by A. D. Lockwood—Improvements in Ring Spindles—Oliver Pearl—Jacob H. Sawyer—George Draper—Richard Garsed—Barton H. Jenks—F. J. Rabbeth—Opening and Picking—Creighton Willow—Richard Kitson—Whitehead & Atherton—Palmer & Jillson—Great Increase at Fall River—New Mills in New England and the Southern States.. 63

CHAPTER XII.

Statistics of 1874—Calico Printing-Machines in 1876—Ginghams—Cottonades—Cotton Duck—Spool-Cotton—Small Wares—Statistics of 1876—Lowell—Lawrence—Manchester—Lewiston—Fall River—Comparative cost of Water and Steam-Power—Water-Wheels at the Centennial—Letter of Edward Atkinson to New York "Herald".. 75

SUPPLEMENTARY CHAPTER.

The Cotton Manufacture as illustrated at the Centennial Exhibition............... 89

APPENDICES.

A.—Paper by William A. Burke, Esq.. 95
B.—Letter of Aza Arnold... 102

TESTS OF POWER.—COTTON-OPENERS,

Date.	Place.	Machine.	No. Beaters.	Rev. do.
May, 1871	Amoskeag Mills, Manches'r, N. H.	Double Creighton Willow,	2	946
" 1872	Masconomet Mill, Newburyport,	Single " "	1	820
" 1871	Amoskeag Mills,	Van Winkle Opener,	1	520
" "	" "	" " "	1	532
Aug., 1871	Manchaug Mill, Mass.	Kitson's "Old Pattern,"	2	1,344
May, 1872	Rockport Mill, "	" " "	2	1,400
" "	" " "	Kitson's Mixer,	1	750
Nov. 1873	Tremont Mills,	Whitehe'd & Atherton,	1	700
" "	Lowell, Mass.,	New Mixer.	1	700

POWER OF COTTON-PICKERS.

Date.	Place.	Machine.	No. Beaters.	Revolution Beaters.
May, 1871	Amoskeag Mills, Manchester, N. H.	36 inch Amoskeag Pattern	3	1,507
"	"	" " "	2	1,026
"	"	Whitin's Pattern,*	3	1,617
June, 1871	Derry Mills, Manch'r	Whitin's Lapper, 30 inch.	2	2,045
May, 1872	Rockport, Mass.,	" " 48 "	3	1,500
April, 1873	"Social" Mill, Woonsocket,	" " 30 "	3	1,500
"	"	" " 30 "	3	2,100
April, 1872	Salmon Falls, N. H.,	Platt's " 36 "	2	1,100
May, 1872	Masconomet, Newburyport,	" " 48 "	2	1,016
"	"	" " 48 "	2	1,066
Nov., 1873	Granite Mill, F. Riv'r	" " 36 "	2	1,130
"	"	" " 36 "	2	1,130
Aug., 1871	Mauchaug, Mass.,	Kitson's "	2	1,344
Jan., 1872	Whittenton, Taunton, Mass.,	" "	2	1,530
Nov., 1872	Essex Mill, Paterson, N. J.,	" "	2	1,066
Nov., 1873	Westville, Taunton,	" "	2	1,500
May, 1873	Manchester Print Works,	Kitson's Compound,	4	1,500
"	Weetamoe Mill, Fall River,	" Lapper,	2	1,500
Oct., 1873	Merrimac Manfg. Co., Lowell,	" Compound,	4	1,600
"	" "	" "	4	1,600
"	"	" "	3	1,600

Note.—Ft. lbs., in all cases in this book, refers to lbs. lifted 1 ft. per second, 550 of which = 1. H. P.
* Built at Amoskeag Shop.

DELIVERING COTTON LOOSE ON FLOOR.

No. Fans.	Rev. do.	Lbs. Cott'n per Day.	Ft. Lbs. per sec.	Horse-Power.	Remarks.
1	1,420	5,000	6,868	12.488	Counter-shaft included { Fan only, 8.024 H. P.
1	1,183	3,000	2,971	5.402	" "
1	1,155	2,000	891	1.620	" "
1	1,360	3,000	1,149	2.090	" "
2	{ 1,456 / 1,620	3,200	3,298	5.996	Cotton blown through long dust-box.
2	1,600	3,000	3,151	5.730	" previously opened in mixer "
1	750	3,000	697	1.258	" delivered to last machine.
1	700	8,330	3,490	6.345	
1	700	10,900	3,679	6.689	

DELIVERING COTTON IN LAP.

No. Fans.	Revolution Fans.	Lbs. Cott'n per Day.	Wt. Lap per Yd.	Ft. Lbs. per Sec.	Horse-Power.	Remarks.
3	1,822	1,000	......	2,670	4.860	1st Picker. 2 Beat's & Fans=2,024=3.68
2	1,200	1,000	......	1,676	3.048	2d " without Feed-motion=2.487
1	1,560	1,000	......	1,622	2.950	2d " " " =2.252
1	2,000	600	......	1,387	2.395	1st "
1	870	1,500	......	2,769	5.034	2d "
2	....	....	......	3,487	6.340	1st " & Hayden Trunk & Dust-box.
1	....	....	......	2,703	4.914	2d "
2	1,600	....	......	2,486	4.520	1st "
2	1,354	....	......	2,667	4.848	1st " without Feed =3.584
2	1,421	....	......	2,511	4.566	2d "
2	1,507	....	10½ oz.	3,237	5.886	1st " without Feed =3.685
2	1,507	....	......	3,441	6.256	2d "
2	1,456	1,200	......	2,080	3.776	1st " without Feed =2.976
2	1,668	....	......	2,514	4.571	2d "
2	1,177	....	......	1,867	3.394	2d " with Evener.
2	1,500	2,000	11 oz.	2,830	5.145	2d " " "
3	{ 2. 1,500 / 1. 2,000	2,000	8 oz.	6,025	10.954	1st "
2	1,500	....	12 oz.	3,045	5.536	2d " with Evener.
3	{ 2. 1,600 / 1. 2,100	3,300	11 oz.	6,807	12.360	1st " on previously-opened Cotton.
3	"	....	......	5,795	10.530	1st " same Machine without Cotton.
2	{ 1. 1,600 / 1. 2,100	....	......	3,897	7.086	1st " " 1 Beater and Fan stopped

COTTON OPENERS AND LAPPERS.

Date.	Place.	Machine.	No. Beaters.	Revolution Beaters.
Oct., 1873	Merrimac Mnfg. Co., Lowell,	Kitson's Compound,	4	1,600
"	"	" "	4	1,600
"	"	" "	4	1,380
"	"	" "	4	1,380
"	"	" "	4	1,380
"	"	" 2d Lapper,	2	1,700
"	"	" "	2	1,700
"	"	" "	2	1,550
"	"	" "	2	1,550
"	"	" "	2	1,400
"	"	" "	2	1,400
Nov., 1873	Tremont Mills, Lowell,	" Compound	4	1,380
"	"	" "	4	1,500
"	"	" "	4	1,500
"	"	Whitehead & Atherton Whipper Lapper,	3	1,300
"	"	" "	3	1,300
"	"	" "	3	1,300
"	"	" "	3	1,300
May, 1873	Great Falls, N. H.,	Whitehead & Atherton Old 1st Lapper,	3	1. 1,200 / 2. 1,500
"	"	" 2d "	2	1,500
Nov., 1873	Westville Mill, Taunton,	Kitson Lapper,	2	1,500
Jan., 1874	Clipper Mill, Baltimore, Md.	" "	3	1,380
"	"	Whitehead & Atherton,	3	1,380
Feb., 1874	Jackson Co. Mills, Nashua,	Kitson Compound,	4	1,380
"	"	" "	4	1,380
"	Boott Mills, Lowell,	Kitson Compound, New Style, with 2 "Broken" Beaters,	4	1. 700 / 1. 950 / 2. 1,380
"	Whittenton Mills, Taunton,	Kitson Compound, Old Style,	4	 1,390
Mar., 1874	"	Same Machine, 1st pair Beaters removed, and 1.24 in. Whipper Cylinder substituted by Whitehead & Atherton, 1,000 rev. p. min.	3	1. 1,000 / 2. 1,390
"	"	1.24 inch "Broken" Beater substituted for Whipper, by Kitson,	3	1. 1,000 / 2. 1,390

COTTON LAPPERS—(Continued).

No. Fans.	Revolution Fans.	Lbs. Cott'n per Day.	Wt. Lap per Yd.	Ft. Lbs. per Sec.	Horse-Power.	Remarks.	
3 {	2. 1,600 1. 2,100	5,000	19¼ oz.	7,333	13.33	1st Trial, A. M., Heavy Lap.	
3 {	2. 1,600 1. 2,100	5,000	19½ oz.	7,414	13.48	2d " P. M., "	
3 {	2. 1.380 1. 1,850	3,000	11 oz.	4,815	8.75	Speed reduced.	
3 {	2. 1,380 1. 1,850	....	....	3,889	7.07	" without Cotton passing.	All on previously-opened Cotton.
3 {	2. 1,380 1. 1,850	4,500	18 oz.	5,864	10.66	Speed same, Lap increased.	
2	1,700	2,250	8¼ oz.	3,744	6.80	"regular, following last machine.	
2	1,700	....	....	2,615	4.755	" " without Cotton.	
2	1,550	2,300	10½ oz.	2,923	5.315	" reduced, Lap heavier.	
2	1,550	....	....	2,141	3.891	" " without Cotton.	
2	1,400	1,750	8¼ oz.	2,410	4.382	" further reduced.	
2	1,400	....	....	1,667	3.080	" " " without Cotton.	
3	1,380	4,200	12¼ oz.	8,871	16.128	2 Trials on Cotton from Bale.	
3	1,500	3,540	11¾ oz.	8,518	15.487	1 Trial " "	
3	1,500	....	....	7,106	12.92	1 Trial without Cotton "	
3	1,300	3,260	13 oz.	4,482	8.15	Taken as running "	
3	1,300	3,600	13¼ oz.	4,623	8.407	" " "	
3	1,300	4,080	14½ oz.	4,448	8.085	" " "	
3	1,300	....	....	3,918	7.123	Without Cotton.	
3 {	2. 1,500 1. 2,000	....	12 oz.	5,116	9.300	Opened Cotton, 1 Scratcher, 2 Beaters.	
2	1,500	....	8 oz.	3,687	6.70	Following last Machine.	
2	1,500	2,000	11 oz.	2,830	5.145	2d Picker.	
3	1,380	3,930	16 oz.	4,383	7.969	1st Picker, working Cotton from Bale.	
3	1,380	3,600	12 oz.	5,883	10.607	" " "	
3	1,380	3,840	13 oz.	5,674	10.318	" " "	
3	1,380	....	....	4,590	8.346	" without Cotton.	
3 {	1. 1,780 2. 1,380	4,800	14 oz.	4,525 3,282	8.228 5.950	" working Cotton from Bale. " without Cotton.	
3	1,390	3,300	15 oz.	8,142 5,047	14.805 9.177	Work'g Bl'k Cotton, dyed after Card'g. Without Cotton.	
3 {	1. 1,000 2. 1,390	.... 3,350 4,420	.... 15 oz. 19½ oz.	3,906 6,308 5,000	7.102 11.505 9.091	" " Work'g Black Cotton, as at above test. Working White Cotton from Bale.	
3 {	1. 1,000 2. 1,390	 3,300	.... 15 oz.	3,691 6,588	6.711 11.978	Without Cotton. Working Dyed Cotton, as before.	

COTTON CARDS.

Date.	Place.	Description.	Width.	Rev'l'n	Lb. Cot. per Day.
June, 1871	Derry Mills, Manchester, N. H.,	Hand-Stripper,	30	116	30
"	Amoskeag Mills, Manches'r, N. H.,	Self-Stripper,	36	110	36
Aug., 1871	Mauchaug, Mass.,	Saco W. P. Co. " Breaker,	36	128	40
"	" "	" " Finisher,	36	128	40
Jan., 1872	Whittenton, Taunton,	Mason's Breaker,	30	127	..
"	" "	" Self-Stripper, Finisher,	36	120	..
"	" "	" " Breaker,	36	120	..
"	" "	" " Single,	36	120	..
Mar., 1872	Haydensville, Mass.,	Whitin's Self-Stripper,	36	120	..
April, 1872	Salmon Falls, N. H.,	Saco W. P. Co. " Breaker,	36	125	..
"	" "	" " Finisher,	36	125	..
May, 1872	Rockport, Mass.,	Whitin's Breaker,	48	137	..
"	" "	Saco W. P. Co. S. S. Finsh'r,	36	123	..
"	Masconomet Mill,	Mason's Breaker,	24	133	..
	Newburyport, Mass.,	Saco W. P. Co. Finisher,	36	127	..
June, 1872	Manchester Print Works, N. H.,	" " "	36	120	27
Nov., 1872	Essex Mill, Paterson, N. J.,	Howard & Bullough,	36	115	76
Apr., 1873	Clinton Mill,	Mason's Cylinder, Wood,	30	130	..
"	Woonsocket, Mass.,	" " Iron,	30	132	..
May, 1873	Weetamoe, F. River,	J. Pettee, Single,	36	120	..
Nov., 1873	Granite, "	Davol & Co.,	36	136	..
June, 1873	Ocean Mill, Newburyport,	Saco W. P. Co. Self-Strip'r,	36	125	..
"	"	" "	36	125	..
"	"	" "	36	125	..
Sept., 1873	Atlantic Mills,	Lowell Ma. Shop, "	36	130	65
"	Lawrence, Mass.,	" "	36	130	65
"	"	" "	36	130	65
"	"	" "	36	130	65
"	"	" "	36	130	45
Nov., 1873	Westville, Taunton,	Mason,	36	156	36

COTTON CARDS—(Continued).

Ft. Lbs. per Sec.	Horse-Power.	No. per Railw'y.	H. P. of Railw'y.	Total H. P.	Cards per H.P.	REMARKS.
44.85	.081*	10	0.585	1.40	7.14	Single Carding for Hosiery, old.
78.92	.144	11	0.645	2.229	5.	" " " Tickings.
46.92	.085	52	1.437	5.857	8.88	2 Tests, Double Carding for fine Cambrics.
70.77	.129	13	.530	2.207	5.89	
98.	.178	28	.806	5.790	4.83	Hand-Stripper, very old.
62.	.112	10	.361	1.481	6.75	
80.	.145	22	1.016	4.206	5.23	
80.	.145	11	.380	1.975	5.57	
70.	.126	9	.253	1.387	6.47	Single Carding.
50.58	.093	64	1.020	6.972	9.18	Double "
50.58	.093	16	.233	1.721	8.26	" "
201.36	.366	48	1.794	5.088	2.48	Hand-Stripper, old.
85.	.155	12	.361	2.221	5.40	Self "
105.	.191	68	2.539	15.527	4.40	Hand-Stripper, Old.
147.	.268	11	.512	3.460	3.18	Self "
40.	.073	12	.601	1.477	8.12	" "
280.	.527	..	....	....	1.90	Single Card, Coiler, 8 W'kers & Strip'rs,
158.	.288	24	.247	7.159	3.35	Hand-Stripper, Breaker, old.
113.21	.206	12	.507	2.979	4.03	Self " , Finisher, "
120.	.218	12	*.667	3.283	3.66	" "
92.	.167	15	.689	3.194	4.70	
63.57	.116	10	.430	1.590	6.29	Single Carding.
66.07	.120	32	1.535	5.375	6.	Breaker.
65.	.118	8	.354	1.298	6.16	Finisher.
150.	.273	60	2.267	18.647	3.28	Breaker.
143.75	.261	60	2.267	17.927	3.35	"
139.13	.253	9	.906	3.183	2.83	Finisher.
125.22	.228	9	.906	2.958	3.04	"
82.92	.151	9	.906	2.265	3.97	"
76.74	.139*	11	.803	2.337	4.71	Single Carding.

* Estimated.

RAILWAY-HEADS FOR CARDS.

Date.		Place.	Description.	
June,	1871	Amoskeag Mills, N. H.,	Breaker, Lap-Head,	
August,	1871	Manchaug Mills, Mass.,	"	"
January,	1872	Whittenton Mills, Mass.,	"	"
"	"	" " "	"	"
May,	"	Salmon Falls, N. H.,	"	"
"	"	Rockport, Mass.,	"	"
"	"	Masconomet, Newburyport, Mass.,	"	"
June,	"	Manchester Print Works, N. H.,	"	"
April,	1873	Clinton Mill, Woonsocket, R. I.,	"	"
June,	"	Ocean Mill, Newburyport, Mass.,	"	"
September,	"	Atlantic Mill, Lawrence, Mass.,	"	"
June,	1871	Derry Mill, Manchester, N. H.,	Finisher Railway,	
"	"	Amoskeag Mills, Manchester, N. H.,	"	"
"	"	" " " "	"	"
August,	"	Manchaug Mills, Mass.,	"	"
January,	1872	Whittenton, Taunton, Mass.,	"	"
March,	"	Haydensville, Mass.,	"	"
"	"	Salmon Falls, N. H.,	"	"
May,	"	Rockport, Mass.,	"	"
"	"	Masconomet, Newburyport, Mass.,	"	"
June,	"	Manchester Print Works, N. H.,	"	"
April,	1873	Clinton Mill, Woonsocket, R. I.,	"	"
June,	"	Ocean Mill, Newburyport, Mass.,	"	"
"	"	" " " "	"	"
September,	"	Atlantic, Lawrence, Mass.,	"	"
November,	"	Granite, Fall River, Mass.,	"	"
"	"	Westville, Taunton, "	"	"
"	"	" "	"	"

RAILWAY-HEADS FOR CARDS.

No. of Cards.	Diameter of Roll.	Velocity of Roll.	Ft. lb. per Sec.	Horse-Power.
32	9 inches.	10 yds. per min.	578.	1.051
52	"	9.42 " "	790.	1.437
28	"	10. " "	443.	.806
22	"	10. " "	559.	1.016
64	"	7.38 " "	560.	1.020
48, 48in	"	10. " "	987.	1.794
68, 24in	"	11. " "	1396.	2.539
36	"	10. " "	519.	.944
24	Can.	11.66 " "	135.76	.247
32	9 inches.	12. " "	844.	1.555
60	5 inches.	14.5 " "	1247.	2.267
10	1½ inches.	320 revolutions.	321.	.585
11	"	400 "	395.	.716
11	"	378 "	361.	.656
13	"	290 "	291.46	.530
10	"	302 "	188.	.361
9	"	220 "	139.	.253
8	"	230 "	128.	.233
12	" .	200 "	198.38	.361
11	"	360 "	282.	.512
12	"	282 "	336.	.601
12	"	394 "	278.57	.507
10	"	200 "	236.36	.430
8	"	200 "	194.44	.354
9	"	365 "	498.	.906
15	"	412 "	379.17	.689
11	"	306 "	310.	.564
11	"	312 "	441.51	.803

DRAWING-FRAMES.—COTTON

DATE.		PLACE.	MAKER.
June,	1871	Derry Mills, Manchester, N. H.,	Unknown,
"	"	Amoskeag Mills, Amoskeag, "	Amoskeag Co., 1st,
"	"	" " " "	" " 2d,
July,	"	Laugdon Mills,	Whitin,
August,	"	Manchaug Mills, Mass.,	Saco W. P. Co., 1st,
"	"	" " "	" " 2d,
January,	1872	Whittenton, Taunton, Mass.,	Mason,
"	"	" " "	"
"	"	" " "	"
March,	"	Haydensville, Mass.,	Whitin,
April,	"	Salmon Falls, N. H.,	Saco W. P. Co.
May,	"	Rockport, Mass.,	Mason,
"	"	Masconomet, Newburyport, Mass.,	" 1st,
"	"	" "	" 2d,
June,	"	Manchester Print Works, N. H.,	Saco W. P. Co. 1st,
"	"	" " " "	" " 2d,
November,	"	Essex, Paterson, N. J.,	Howard & Bullough,
April,	1873	Clinton, Woonsocket, R. I.,	Whitin, 1st,
		" " "	" 2d,
		" " "	" 3d,
May,	"	Weetamoe, Fall River, Mass.,	Whitin, 1st,
		" " "	" 2d,
June,	"	Ocean, Newburyport, "	Higgins, 1st,
"	"	" "	" 2d,
November,	"	Granite, Fall River, "	Whitin, 1st,
"	"	" "	" 2d,
"	"	Westville, Taunton, "	Mason, 1st,
		" " "	" 2d,

DRAWING-FRAMES.—COTTON.

No. Rolls.	Rev. per min.	Doublings.	Draft.	From	To	No. Deliveries.	Ft. Lb. per Delivery.	H.P. per Deliv'ry	H. P. Frame.
3	310	2	4.50	....	....	6	46.20	.084	.506
4	240	2	3.	97.g.	74.g	8	39.4	.072	.573
4	240	4	3.50	74.	80.	8	40.7	.074	.591
4	221	2	4.	79.	40.	6	52.8	.096	.580
5	226	3	4.07	115.	87.	6	60.5	.110	.662
5	226	3	4.30	87.	66.	8	57.75	.105	.842
3	400	2	5.	62.5	12.5	10	47.4	.086	.860
4	340	2	....	....	....	4	105.	.190	.762
4	155	3	3.50	....	....	6	66.6	.121	.727
4	220	4	....	....	....	4	65.	.119	.474
5	196	3	3.34	109.	98.	8	45.5	.083	.662
4	202	4	....	....	....	8	42.73	.078	.621
5	258	3	4.83	123.	75.	6	79.40	.144	.866
5	296	3	4.68	75.	48.	8	75.	.136	1.091
5	184	3	3.33	....	....	12	72.4	.132	1.580
..	240	3	3.75	....	....	12	80.3	.146	1.753
4	338	8	....	....	....	6	74.72	.136	.815
4	381	4	....	....	....	6	104.36	.190	1.138
4	361	4	....	....	....	8	113.75	.207	1.655
4	380	2	....	....	....	20	58.27	.097	1.937
4	220	3	....	....	....	6	72.	.131	.784
4	210	2	....	....	....	12	42.65	.078	.930
4	220	3	....	....	....	12	46.38	.086	1.012
4	220	3	....	....	....	16	38.	.068	1.091
4	238	3	4.	136.	102.	8	63.5	.115	.924
4	238	3	4.61	102.	76.5	12	59.4	.108	1.296
4	312	3	4.50	135.	90.	4	73.21	.134	.534
4	415	3	4.50	90.	55.	4	89.12	.162	.648

DEAD-SPINDLE ROVING-FRAMES.

Date.	Place.	Description.	Size of Bobbin.	No. Spin.	Rev. Spin.
April, 1871	Stark Mills,	Lowell Speeder,	$10^2 \times 5^3$	28	720
" "	Manchester, N. H.,	" "	8×4	52	904
" "	" "	" "	"	64	830
June, "	Amoskeag Mills,	" " built	12×6	30	501
" "	Manchester,	by Amoskeag Co.,	10×5	40	601
" "	"	"	"	40	575
" "	"	"	8×4	46	887
" "	"	"	"	64	782
" "	"	"	"	64	906
Mar., 1872	Haydensville,	Brown Speeder,	9×4^5	30	....
" "	"	" "	7×3^2	44	....
" 1873	Am. Linen Co., F. River,	" "	"	78	1277
Nov., "	Westville, Taunton,	" "	"	60	962
Mar., 1872	Haydensville,	Pettee, "Soft Bobbin,"	"	40	....

ROVING-FRAMES.

Date.	Place.	Description.	Size of Bobbin.	No. Spin.	Rev. Spin.
Aug., 1871	Manchaug, Mass.,	Saco W. P. Co. Slubber	12×6	48	590
Sept., "	Wauregan, Conn.,	" "	"	52	540
" "	" "	Higgins & Sons' "	"	40	475
April, 1872	Salmon Falls, N. H.,	" "	"	56	530
May, "	Masconomet, Newburyport, Mass..	Saco W. P. Co. "	"	60	543
June, "	Manchester P. Works,	" "	"	44	360
Mar., 1873	Am. Linen Co., F. River,	Higgins & Sons' "	"	64	606
" "	" " "	" "	"	60	515
April, "	Mannville, R. I.,	Walker & Hacking "	"	64	543
May, "	Wectamoe, Fall River,	Prov. Mach. Co. "	"	68	648
June, "	Ocean, Newburyport, Mass.,	Curtis, Parr & Co. "	"	48	550
" "		Higgins & Sons' "	"	48	560
Nov., "	Granite Mills,	" "	"	68	560
" "	Fall River, Mass.,	How'd & Bullough "	"	48	630
		" "	"	56	630

DEAD-SPINDLE ROVING-FRAMES.

Diam. of Roll.	Rev. Roll.	Draft.	From	To	Ft. Lb. Frame.	Ft. Lb. Spindle.	H. P. Frame.	Spindl's H. P.	REMARKS.
$1\frac{1}{4}$	175	4.	72.gr.	18.3g.	465	16.6	.845	33.07	Average half-full Bobbin.
$1\frac{1}{8}$	162	6.	18.3	5.$\frac{1}{8}$	597	11.49	1.086	47.88	"
"	149	6.	18.3	5.$\frac{1}{8}$	626	9.78	1.138	56.24	"
$1\frac{1}{4}$	178	3.50	80.	25.	461	15.36	.838	35.80	"
$1\frac{3}{16}$	170	4.	25.	12.50	522	13.05	.949	41.15	"
"	159	4.24	75.	17.70	554	13.85	1.007	39.72	"
$1\frac{1}{8}$	180	5.78	17.70	1.38h	682	14.83	1.240	37.10	"
"	126	6.04	12.50	2. hk.	563	9.06	1.023	62.56	"
"	144	6.04	12.50	2. hk.	670	10.47	1.218	52.54	"
$1\frac{1}{4}$	....	....	....	......	585	19.50	1.064	28.20	"
$1\frac{1}{8}$	....	....	...	......	1,037	23.58	1.887	23.32	"
"	125	6.60	1.25hk	4.13 h	859	11.01	1.561	50.	"
"	175	6.50	55.gr.	0.95 h	830	13.63	1.509	40.	"
"	....	...	....	......	500	12.50	.910	44.	"

ROVING-FRAMES.

Diam. of Roll.	Rev. Roll.	Draft.	From	To	Ft. Lb. Frame.	Ft. Lb. Spindle.	H. P. Frame.	Spindl's H. P.	REMARKS.
$1\frac{1}{4}$	152	5.63	66.gr.	0.52 h	739	15.39	1.343	35.	Average half full.
"	135	4.79	72.85	15.21 g	684	13.15	1.244	41.8	" "
"	125	4.48	60.	13.	521.20	13.03	.948	42.13	" "
"	142	4.72	85.	18.	692.	12.36	1.259	44.4	" New Frame.
"	183	3.80	47.9	0.56 h	796.5	13.27	1.448	41.4	"
"	92	4.03	0.13hk	0.53 h	357	8.12	.650	68.	"
"	212	....	......	0.43	1220	19.06	2.217	28.75	" Old Frame.
"	180	....	......	0.52	1154	19.24	2.100	28.	" "
"	190	....	... ..	0.49	778.83	12.16	1.419	45.	"
"	107	4.	0.14	0.55	621	9.13	1.129	60.	"
"	156	....	......	......	738.5	15.38	.738.5	35.76	"
$1\frac{1}{8}$	150	4.17	0.12	0.50	454	9.44	.824	58.25	"
"	150	4.17	0.12	0.50	827	12.16	1.503	45.50	"
$1\frac{1}{4}$	174	4.53	76.5gr.	0.39	725	15.11	1.318	36.42	"
"	174	4.53	76.5gr.	0.39	862	15.51	1.567	35.75	"

ROVING-FRAMES.—(*Continued.*)

Date.	Place.	Description.	Size of Bobbin.	No. Spin.	Rev. Spin.
July, 1871	Langdon Mills, N. H.	Prov. Ma. Co. Slubber,	10 × 5	48	615
Aug., "	Manchaug Mills, Mass.	Saco W. P. Co. Inter.,	"	80	736
" "	" " "	City M. Co. Slubber,	"	88	720
Sept., "	Wauregan, Conn.,	Higgins & Sons "	"	52	509
Jan., 1872	{ Whittenton Mills,	Wm. Mason "	"	56	670
" "	{ Taunton, Mass.	" "	"	64	634
April, "	Salmon Falls, N. H.,	Higgins & Sons Inter.,	"	88	630
Nov., "	Essex, Paterson, N. J.,	How'd & Bull'h Slub.,	"	80	564
" "	" " "	Saco W. P. Co. "	"	60	525
Dec., "	Wash'gton, Gloster,N. J.,	Higgins & Sons "	"	76	750
Jan., 1873	Potomska, N. Bedford,	City Ma. Co. "	"	52	450
April, "	Manville, R. I.,	Prov. Ma. Co. 1st In.,	"	80	769
" "	Social, Woonsocket, R. I.	City Ma. Co. Slubber,	"	80	714
May, "	Weotamoe, Fall River,	Curtis, Parr & Co. In.,	"	68	650
June, "	{ Ocean Mills, New-	Higgins & Sons In- }	"	64	780
" "	{ buryport, Mass.,	termediates. }	"	88	780
Nov., "	Granite, Fall River,	How'd & Bull'h Inter.,	"	66	694
May, 1872	Rockport, Mass.,	Prov. Ma. Co. Slubber,	"	60	477

ROVING-FRAMES.

Date.	Place.	Description.	Size of Bobbin.	No. Spin.	Rev. Spin.
July, 1871	Amoskeag, Manchester,	Prov. Ma. Co. Interm.,	9 × 4²	84	815
" "	Langdon, "	" "	"	80	773
Sept., "	Wauregan, Conn.,	Higgins & Sons "	"	96	575
Jan., 1872	Whittenton, Taunton,	Mason "	"	72	676
May, "	Rockport, Mass.,	Prov. Ma. Co. "	"	72	588
" "		Saco W. P. Co. "	"	84	604
" "	Masconomet, Mass.,	Prov. Ma. Co. "	"	80	630
June, "	Manchester P. W., N. H.	SacoW. P. Co. 1st "	"	72	550
Nov., "	Essex, Paterson, N. J.,	How'd & Bull'h "	"	100	562
Mar., 1873	{ Am. Linen Co., Fall	Higgins & Sons "	"	72	730
" "	{ River, Mass.,	Walker & Hacking"	"	72	781
April, "	Clinton, Woonsocket,	Higgins & Sons Slub.,	"	72	530
Nov., "	{ Westville, Taunton,	Prov. Ma. Co. "	"	80	880
" "	{ Mass.,	" "	"	80	576
Sept., 1871	Wauregan, Conn.,	Higgins & Sons Inter.,	8 × 4	80	709
Dec., 1872	Washington, N. J.,	" "	"	96	900
Jan., 1873	Potomska, N. Bedford,	City Mach. Co. "	"	104	575
April, "	Manville, R. I.,	Prov. Ma. Co. 2d "	"	136	952

25

ROVING-FRAMES.—(Continued.)

Diam. of Roll.	Rev. Roll.	Draft.	From	To	Ft. Lb. Frame.	Ft. Lb. Spindle.	H. P. Frame.	Spindl's H. P.	REMARKS.
1¼	138	4.	39.77g	0.84h	434.	9.04	.789	60.86	Half-fall Bobbin.
"	133	5.03	0.52h	1.25	676.	8.98	1.300	62.	" "
"	216	4.03	7.¼g	0.54	1550	17.62	2.819	31.21	" "
"	130	4.03	......	....	554.	10.66	1.008	51,67	" "
"	120	4.80	12½.g.	0.66	747.	13.33	1.358	41.26	" "
"	173	4.50	......	....	925.	14.45	1.682	38.05	" "
"	112	5.14	18.g.	7.g.	600.	6.82	1.091	80.66	" "
1½	176	6.	70.25	0.69	879.	10.99	1.595	50.	" "
1 3/16	101	6.	50.g.	1.00	470.6	7.84	.856	70.	" "
....	200	....	......	....	1650	22.71	3.000	25.30	" "
1¼	130	4.	0.13h	0.55	446.	8.58	.811	64.	" "
"	95	4.2	0.55	1.30	577.	7.21	1.049	76.	" "
"	120	5.41	......	....	848.	10.60	1.541	52.	" "
1½	140	5.30	0.50	1.32	939.	13.81	1.707	40.	" "
"	130	5.30	0.50	1.32	483.76	7.56	.878	73.	" "
"	130	5.30	0.50	1.32	996.	11.32	1.811	48.5	" New Frame
1¼	124	6.20	0.39	1.21	780.	11.82	1.418	46.5	" just started.
"	145	4.90	......	....	469.	7.81	.853	70.	"

ROVING-FRAMES.—(Continued.)

Diam. of Roll.	Rev. Roll.	Draft.	From	To	Ft. Lb. Frame.	Ft. Lb. Spindle.	H. P. Frame.	Spindl's H. P.	REMARKS.
1¼	140	5.25	0.47h	1.25h	575	6.84	1.045	80.58	Half-fall Bobbin.
"	118	4.50	0.84	1.88	681	8.51	1.238	64.62	" "
"	93	5.07	13.g.	1.50	723	7.53	1.314	73.06	" "
"	120	....	....	....	814	11.30	1.480	48.65	" "
"	100	5.20	....	....	475	6.60	.863	83.	" "
"	108	5.20	....	....	507	6.04	.924	91.	" "
"	100	5.27	0.56	1.49	537	6.72	.977	82.	" "
"	110	5.03	0.53	1.30	417	5.79	.758	95.	" "
1½	107	4.20	0.69	1.45	781	7.80	1.420	70.	" "
1½	138	5.50	0.43	1.20	911	12.05	1.656	43.5	" "
"	140	5.	0.49	1.26	858	11.92	1.561	46.	" "
"	118	4.80	....	1.14	326	4.59	.593	120.	" Centrifugal
"	174	6.50	55.g.	9.95	937	11.14	1.705	47.	" Presser.
"	141	3.76	55.g.	0.55	833	9.47	1.585	58.	"
1¼	125	5.67	0.85	2.33	539	6.74	.981	81.5	"
"	180	....	....	....	1,175	12.24	2.137	45.	"
1½	118	6.18	0.55	1.39	573	5.51	1.042	100.	"
"	100	5.	1.30	3.27	885	6.53	1.610	84.	"

ROVING-FRAMES.—(*Continued.*)

Date.	Place.	Description.	Size of Bobbin.	No. Spin.	Rev. Spin.
July, 1871	Langdon Mills, Manchester, N. H.	Prov. Machine Co. Fine F. Frame,	7 × 3½	144	934
Aug., "	Manchaug, Mass.,	Saco W. P. Co. 2d In.,	"	136	968
" "	" " "	" Fine F. Frame,	"	136	979
Sept., "	Wauregan, Conn.,	" "	"	128	993
" "	" "	Higgins & Sons "	"	128	993
Jan., 1872	Whittenton Mills,	Wm. Mason "	"	112	779
" "	Taunton, Mass.,	" "	"	128	935
" "	" "	" "	"	112	935
" "	" "	" "	"	136	861
April, "	Salmon Falls, N. H.,	Higgins & Sons "	"	152	1060
May, "	Rockport, Mass.,	Saco W. P. Co., "	"	128	904
June, "	Manchester P. W., N. H.	" 2d Interme.,	"	144	1000
Nov., "	Essex, Paterson, N. J.	How'd & Bull. F. F. F.	"	144	873
" "	" " "	Saco W. P. Co. "	"	144	775
Dec., "	Wash'gton, Gloucst., N.J.	Higgins & Sons "	7 × 3	140	1350
Mar., 1873	Am. Linen Co., F. River,	" "	7 × 3½	144	1305
April, "	Clinton Mill, Woonsocket, Mass.,	" "	"	120	760
" "	" "	" "	"	120	744
" "	" "	Prov. Mach. Co. "	"	120	707
" "	Social Mill, "	City " "	"	160	1041
" "	" "	" " "	"	160	1041
" "	" "	Wm. Mason "	"	160	900
May, "	Weetamoc, Fall River,	Curtis, Parr & Co. "	"	144	1060
June, "	Ocean Mill, Newburyport, Mass.,	Higgins & Sons "	"	136	1160
" "	" "	" "	"	152	1160
Nov., "	Granite Mill, Fall River,	How'd & Bull'h, "	"	160	1070
July, 1871	Amoskeag, Man'r, N. H.,	Prov. M. Co. F. F. F.,	6 × 3	128	611
May, 1872	Rockport, Mass.,	" "	"	128	900
" "	Masconomet, New'port,	" "	"	136	1000
Jan., 1873	Potomska, N. Bedford,	City Mach. Co. "	"	168	900
April, "	Manville, R. I.,	Prov. " " "	"	184	1129
Nov., "	Westville, Taunton,	" " " "	"	128	832
Aug., 1875	Manchaug, Mass.,	Saco W. P. Co. F.F.F.,	5 × 2½	144	1117
June, 1871	Manchester P.W., N. H.,	" " "	"	152	1340

ROVING-FRAMES.—(*Continued.*)

Diam. of Roll.	Rev. Roll.	Draft.	From	To	Ft. Lb. Frame.	Ft. Lb. Spindle.	H. P. Frame.	Spindles per H.P.	REMARKS.
1⅛	98	5.	1.88 h.	4.76 h.	577	4.01	1.050	137.	Bob. ¼ full.
"	115	5.90	1.25	4.	823	6.05	1.496	91.	" "
"	94	6.30	1.87	5.95	716	5.26	1.302	104.5	" "
"	106	5.98	1.54	4.75	910	7.11	1.655	77.4	" "
"	106	5.98	1.54	4.75	1,093	8.54	1.988	64.4	" "
"	130	6.	0.66	1.90	645	5.76	1.173	97.	" " Old Pattern
"	104	4.54	0.88	4.	692	5.41	1.250	102.	" " New "
"	104	4.54	0.88	4.	584	5.22	1.063	105.	" " " "
"	133	6.	0.66	1.90	757	5.57	1.377	99.	" " Col'd Rov.
"	110	6.	1.21	3.62	691	4.55	1.256	121.	" " New Pat'rn
"	98	6.	1.32	4.	512	4.	.931	137.	" "
"	100	5.45	1.30	3.64	646	4.49	1.175	122.5	" "
"	97	5.34	1.49	4.	733	5.09	1.333	108.	" "
"	94	6.66	1.	3.33	873	6.06	1.588	91.	" "
"	141	....	....	4.	1,534	10.96	2.789	50.	" "
"	123	6.60	1.25	4.13	1,046	7.27	1.903	76.	" "
"	95	6.60	1.14	3.80	549	4.57	.998	120.	" "Cent. Pres'r
"	93	6.60	1.14	3.80	506	4.22	.920	130.	" " Spring "
"	86	6.60	1.14	3.80	413	3.44	.733	164.	" " " "
"	120	6.86	1.27	4.37	1,167	7.28	2.121	75.2	" " Bad order.
"	120	6.86	1.27	4.37	900	5.63	1.635	98.	" " Sim. Frame
"	97	6.86	1.27	4.37	622	3.87	1.131	141.	" "
1 in.	120	6.40	1.20	3.80	995	6.91	1.808	80.	" "
"	120	6.	1.32	3.90	842	6.19	1.540	83.	" "
"	120	6.	1.32	3.90	1,065	7.	1.937	79.	" " New Frame
1¼	126	6.40	1.21	3.88	1,414	8.83	2.570	62.25	" "
"	89	5.92	1.25	3.70	256	2.00	.463	276.	" "
"	98	5.87	....	....	682	5.33	1.241	103.	" " Old Frame.
"	100	6.17	1.49	4.50	541	4.00	.983	138.	" "
"	92	6.10	1.39	4.28	685	4.08	1.245	135.	" "
"	86	5.	3.87	9.	604	3.77	1.263	146.	" "
"	125	7.	0.55	1.95	597	4.67	1.086	118.	" "
"	61	6.	4.	12.	603	4.18	1.096	131.4	" "
"	86	6.	3.64	10.75	631	4.15	1.147	132.5	" "

THROSTLE-SPINNING.

Date.	Place.	Description.	No. Spin.	Weight Flier.	Rev. of Flier.	Rev. Roll.	Draft.
April, 1871	Stark Mills, Man'r, N. H.,	Amosk'g Sh'p 23 years old,	128	3¾ oz.	3,820	94	8.27
" "	"	"	"	"	3,820	94	8.27
" "	' "	Locks and Canals Co., 30 y.	"	4 oz.	4,000	94	7.84
Nov., "	"	Amosk'g Sh'p 23 years,	"	3¼ oz.	4,100	100	8.27
" "	"	"	"	"	4,220	103	8.27
" "	"	"	"	"	4,220	...	...
" "	"	"	"	"	3,690	90	8.27
" "	"	"	"	"	3,280	80	8.27
" "	"	"	"	"	2,954	72	8.27
" "	"	"	"	"	2,685	66	8.27
June, 1872	Amosk'g Mill No. 2	Amosk'g Sh'p about 20 yrs.,	"	4 oz.	3,071	76	9.35
" "	" " 1	"	"	"	3,170	74	10.26
" "	" " 1	"	"	"	3,042	73	9.35
" "	" " 1	"	"	"	3,332	78	9.35
" "	" " 1	"	"	3½ oz.	3,226	76	9.35
" "	" " 3	"	"	4 oz.	3,694	96	6.65
" "	" " 3	"	"	"	3,450	90	6.65
Aug. 7, "	" " 3	"	"	"	3,450	90	6.65
" 8, "	" " 3	"	"	"	3,450	90	6.65
June, "	" " 3	"	"	"	3,690	76	7.09
" "	" " 3	"	"	"	3,820	79	7.09
" "	" " 3	"	"	"	3,717	73	8.27
July, "	" " 5	"	160	2½ oz.	4,136	72	8.39
" "	" " 5	"	"	"	4,142	64	9.93
April, "	Appleton Mills, Lowell,	Lowell M. S'p., old,	128	3⅓ oz.	3,800	81	8.
July, "	Pepperell Mills,	"	"	3.45 oz	4,929	71	7.
" "	Biddeford, Maine,	"	"	"	4,929	71	7.
" "	"	"	"	"	4,929	71	7.
" "	"	"	"	2.83 oz	4,929	71	7.
" "	"	"	"	"	4,929	71	7.
" "	"	"	"	"	4,929	71	7.
Feb., 1874	Nashua Mfg. Co.	"	"	4 oz.	3,778	59	8.66

DEAD-SPINDLE.

From	To	Ft. Lb. Frame.	Ft. Lb. Spindle.	H. P. Frame.	Spindl's H. P.	Remarks.
5⅓ gr.	13. weft.	954	7.45	1.734	74.	Heart Motion on Lift, A.
1.61 h.	13. "	929	7.25	1.689	76.	Mangle " " Similar Frame, B.
1.61	12.33 wn.	1,077	8.38	1.959	65.	
1.61	13. weft.	930	7.25	1.690	76.	Similar to Frame B.
1.61	13. "	867	6.72	1.576	81.	Frame A changed to [Mangle, leveled. [and oiled.
...	......	689	5.38	1.252	102.	" Rolls stopped.
1.61	13. weft.	722	5.64	1.313	97.	" speed reduced, in full operation.
1.61	13. "	604	4.72	1.079	110.	" " " further.
1.61	13. "	486	3.80	.883	145.	" " " "
1.61	13. "	428	3.34	.778	164.	" " " "
1.50	14. "	746	5.83	1.357	94.	
1.50	16. "	688	5.36	1.250	102.	
1.50	14. "	767	6.00	1.395	92.	
1.50	14. "	848	6.62	1.542	83.	Speed increased from last Trial.
1.50	14. "	757	5.91	1.577	93.	Similar Frame, Malleable-iron Flier.
1.50	10. warp.	920	7.19	1.673	76.	Old Frame, refitted in New Mill.
1.50	10. "	881	6.89	1.601	80.	" Speed reduced.
1.50	10. "	757	5.91	1.377	93.	" very hot afternoon. [in night.
1.50	10. "	938	7.33	1.703	75.	" Rainy morning, weather changed
2.	14. "	649	5.07	1.180	108.	Similar Frame. Fair day, 2 Trials, ex-
2.	14. "	740	5.76	1.345	95.	Similar Frame. [ceptionally light.
2.	16. "	739	5.76	1.343	95.	" "
2.86	24. "	971	6.07	1.766	90.	
2.86	28. "	954	5.96	1.735	92.	
1.65	12.50 "	977	7.63	1.776	72.	Banding Tests. [16½ on one side.
3.14	22. "	1,102	8.60	2.003	64.	Com'n Flier, empty Bobbin, long Band,
3.14	22. "	1,058	8.27	1.924	66.5	" " 16 sp. across, 8 each side.
3.14	22. "	1,014	7.92	1.845	69.	" " 8 " 4 "
3.14	22. "	927	7.24	1.686	76.	Pearls Flier 16 on one side.
3.14	22. "	884	6.90	1.607	79.75	" " 16 across, 8 each side.
3.14	22. "	841	6.57	1.528	83.75	" " 8 " 4 "
2.50	22. "	986	7.70	1.793	71.10	Bands hard and heavy.

Date.	Place.	Maker.	Diam. Ring.	No. Spin.	Wt. Spin.	Rev. Spindle.	Rev. Roll.	Draft
April, 1871	Stark Mills, Man'r, N. H.,	Lowell Ma. Shop,	$1\frac{3}{4}$ in	144	$12\frac{1}{4}$ oz	4,480	100.	7.84
Nov., 1871	"	"	"	144	"	4,480	100.	7.84
"	"	"	"	144	"	4,480	....	....
May, 1871	Amosk'g Mills, "	Amoskeag Co.,	$1\frac{1}{2}$	144	$11\frac{1}{2}$	4,380	73.	7.26
"	"	"	"	144	"	5,100	72.	8.48
"	"	"	"	144	"	5,727	75.	7.76
June, 1871	"	"	"	128	13	5,066	62.	7.68
"	"	"	"	128	"	5,240	68.	8.40
July, 1871	Langdon Mills, "	Saco W. P. Co.	"	128	$12\frac{1}{2}$	5,900	$62.^{6}$	7.44
Aug., 1871	Manchaug, Mass.	"	"	160	10	5,857	54.	7.68
Sept., 1871	Wauregan, Conn.,	Whitin,	"	144	12	5,028	63.	7.26
"	"	"	"	144	$12\frac{1}{2}$	4,968	54.	6.75
"	"	"	"	144	"	5,060	53.	6.75
"	"	"	$1\frac{7}{16}$	240	"	4,930	53.	6.75
"	"	"	"	240	"	5,010	64.	6.75
June, 1871	Derry Mills, Man'r, N. H.,	"	2	120	"	2,211	90.	6.
Nov., 1871	Stark Mills, N.H.,	Lowell Ma. Shop,	$1\frac{3}{4}$	144	$12\frac{1}{4}$	4,524	101.	7.84
"	"	"	"	144	"	4,255	95.	7.84
"	"	"	"	144	"	3,893	87.	7.84
"	"	"	"	144	"	3,476	78.	7.84
"	"	"	"	144	"	3,200	73.	7.84
Mar., 1872	Haydensville, Ms.	Whitin,	$1\frac{7}{16}$	120	....	5,609	84.	....
Jan., 1872	Whittenton, Taun'n, Mass.	William Mason,	$1\frac{13}{16}$	128	12	4,325	100.	9.50
"	" No. 1 Mill,	"	"	128	"	4,325	100.	9.50
"	" "	"	"	128	"	3,500	120.	5.
"	" "	"	"	128	"	3,800	100.	7.37
"	" "	"	"	128	"	4,826	114.	8.
"	" "	"	"	128	"	3,800	100.	8.
"	" "	"	"	128	"	3,400	86.	7.37
"	" No. 2 Mill,	"	$1\frac{5}{8}$	160	"	4,050	130.	8.
"	" "	"	"	160	"	5,067	96.	9.50
"	" "	"	"	160	"	5,067	84.	6.50
Mar., 1872	Salmon F'ls, N.H.	Saco W. P Co.,	"	144	11	4,972	68.	7.25
Apr., 1872	Appleton Co., Lowell,	Lowell Ma. Shop,	"	144	12	5,120	100.	8.
May, 1872	Rockport, Mass.,	William Mason,	"	72	"	5,240	68.	7.27
"	"	"	"	108	"	5,440	68.	7.27
"	"	Saco W. P. Co.,	"	192	11	5,460	68.	7.27
"	Masconomet, Newburyp'rt,	"	"	192	"	6,000	75.	6.66
June, 1872	Amoskeag, No. 6,	Amoskeag Co.,	$1\frac{1}{2}$	128	....	5,154	61.	7.68
"	"	"	"	128	....	6,187	68.	7.68
"	"	"	"	128	....	3,544	$74.^{5}$	....
"	Manches'r P. W.,	Saco W. P. Co.,	$1\frac{5}{8}$	128	....	5,950	64.	....
"	Manchester, N.H.	"	"	192	....	5,950	56.	....

RING SPINNING.

From	To	Ft. Lb. Frame.	Ft. Lb. Spin.	H. P. Frame.	Spindl's H. P.	REMARKS.
1.61	12.33 wp	750	5.21	1.363	106.	Old Bands, Bobbin half full.
1.61	12.33 "	865	6.	1.572	92.	Same Frame, new Bands.
....	......	630	4.37	1.145	126.	" Rolls stopped, Spindles only.
3.30	24. "	631	4.38	1.148	126.	
3.30	28. "	715	5.	1.300	110.	Same Frame as last trial. Speed and Draft
4.40	34. "	845	5.87	1.537	94.	Similar ' " [changed·
4.42	34. weft	727	5.68	1.322	97.	
4.42	37. "	794	6.20	1.444	90.	Similar to last Frame.
4.26	33. warp	698	5.45	1.268	101.	
12.	46. "	596	3.73	1.084	148.	Three Trials, averaged. [started that A. M.
4.33	30. "	1,306	9.	2.375	61.	Wet day, Frame near open door, not previously
7.	48. "	1,111	7.70	2.020	71.	Common Bobbin, 1-8 inch Bands.
7.	48. "	979	6.80	1.781	81.	Chambered Bobbin, 1-16 inch Mule Bands.
7.	48. "	1,476	6.15	2.684	90.	Common Bobbin and Band, Draper's Spiral
4.50	30. "	1,451	6.04	2.639	91.	" " " " [Bolster.
1.33	8. "	634	5.28	1.153	104.	Hosiery Frame.
1.61	12.33 "	841	5.77	1.510	95.	
1.61	12.33 "	769	5.34	1.399	103.	Same Frame as last, Speed reduced.
1.61	12.33 "	677	4.70	1.230	117.	" " " further.
1.61	12.33 "	584	4.05	1.062	136.	" " " "
1.61	12.33 "	527	3.66	.959	150.	" " " "
....	23. "	782	6.52	1.423	84.4	
2 {3. 4.	16. "	787	6.15	1.431	89.5	Colored Rovings, Black and Orange.
1.90	16. "	727	5.68	1.322	97.5	White "
1.90	9. weft	677	5.29	1.231	104.	" "
1.90	14. "	649	5.10	1.180	108.	" "
2 {2.50 3.90	11. warp	820	6.40	1.491	86.	Colored "
"	11. "	700	5.46	1.273	100.	" " Black and White.
1.90	14. weft	680	5.31	1.237	103.	White "
1.15	9. warp	1,018	6.36	1.852	86.5	" "
2 {3. 4.	17. "	1,060	6.60	1.927	83.	Colored " Black and Orange.
3.40	22. "	1,055	6.59	1.917	83.5	White "
3.62	25. "	735	5.10	1.336	108.	
1.65	12.50 "	740	5.14	1.346	107.	Average of four Tests.
3.80	28. "	501	6.96	.911	79.	Old Frame.
3.80	28. "	728	6.74	1.323	82.	" "
3.80	28. "	1,023	5.33	1.861	103.	
4.50	30. "	1,179	6.14	2.143	90.	
4.42	34. "	660	5.15	1.200	107.	
4.42	34. "	789	6.16	1.434	89.	Same Frame, last Trial, increased Speed.
....	14. "	480	3.75	.872	147.	
....	30. "	886	6.92	1.611	80.	
....	42. "	1.150	6.	2.091	92.	

RING SPINNING.—(Continued.)

Date.	Place.	Maker.	Diam. Ring.	No. Spin.	Wt. Spin.	Rev. Spindle.	Rev. Roll.	Draft
July, 1872	Pepperell Mills,	Saco W. P. Co.,	1½	128	11.²oz	5,666	84	7.
"	Biddeford, Me.,	"	"	128	"	5,666	84	7.
"	"	"	"	128	6.²	5,666	84	7.
"	"	"	"	128	11.²	5,666	84	7.
"	"	"	"	128	"	5,666	84	7.
"	"	"	"	128	"	5,666	84	7.
Oct., 1872	Amoskeag Mill,	Amoskeag Co.,	1½	144	"	4,600	70	7.26
"	Manchester, N.H.	"	"	144	"	5,250	77	7.26
"	"	"	"	144	"	6,127	90	7.26
"	"	"	"	144	"	7,355	107	7.26
"	"	"	"	128	13.	3,090	66	9.35
"	"	"	"	128	"	4,050	82	9.35
"	"	"	"	128	"	5,672	115	9.35
Dec., 1872	Washington Mill,	Gloucester Foun.,	1 7/16	128	12.	6,000	70	7.30
"	Gloucester, N. J.,	Whitin,	1½	128	"	5,890	63	7.30
"	"	"	"	128	"	6,356	70	7.30
"	"	Lanphear,	"	128	....	6,000	72	7.30
"	"	Fales & Jenks,	"	128	....	6,000	72	7.30
Feb., 1873	{ Cocheco, Dover, N. H.,	Cocheco Co.,	1¾	128	....	4,820	66	7.90
"	"	Saco W. P. Co.,	1⅝	192	....	4,895	67	7.90
"	"	"	"	192	....	4,895	67	7.90
April, 1873	Clinton Mill,	Whitin,	"	192	12.	5,460	68	8.14
"	Woonsocket, R. I.	"	"	192	"	5,540	69	8.14
"	"	"	"	128	"	5,380	67	8.14
"	"	"	"	128	"	6,020	75	8.14
"	Social Mill, R. I.,	"	"	192	"	5,715	64	7.52
"	"	"	"	192	"	5,715	64	....
May, 1873	{ Am. Linen Co., Fall River,	Higgins & Sons,	1½	224	12.⁸³	5,200	56	7.09
"	Davol Mills, "	Saco W. P. Co.,	"	128	11.	6,000	74	6.52
"	"	"	"	128	"	5,270	65	6.52
"	"	"	"	128	"	5,430	67	6.52
June, 1871	Pacific Mills,	Whitin,	1⅝	160	12.	6,059	73	6.79
"	Lawrence, Mass.,	"	"	160	"	6,059	73	6.79
"	"	"	"	160	"	6,059	....	....
"	"	"	"	160	"	6,059	....	....
"	"	" Altered,*	"	160	7.	6,059	73	6.79
"	"	" "	"	160	"	6,059	73	6.79
"	"	" "	"	160	"	6,059	....	....
"	"	" "	"	160	"	6,059	....	....
Sept., 1873	Atlantic Mills,	Lowell Ma. Shop,	1⅝	176	12.	5,802	96	7.60
"	Lawrence, Mass.,	"	"	176	"	5,802	96	7.60
Nov., 1873	Westville, Taunton, Mass.	William Mason, Light Spindle,	"	192	8.	5,864	107	9.45
"			"	192	"	5,651	102	9.45
Mar., 1873	{ Crescent Mill, Fall River,	"	1⅜	192	6.	5,570	67	7.50

* Spindle cut off at butt, and reduced in diameter—top as before.

RING SPINNING.—(*Continued.*)

From	To	Ft. Lb. Frame.	Ft. Lb. Spin.	H. P. Frame.	Spindl's H. P.	Remarks.
3.14	22. warp.	850	6.64	1,545	83.	Spindles banded singly.
3.14	22. "	983	7.69	1.788	71.5	Empty Bobbin } Averages 1.667 H. P., 7.26 lb. per Full " } Spindle, or 77 Spindles per H. P.
3.14	22. "	900	7.03	1.686	78.	Spindle shortened at butt. Average one-half full.
3.14	22. "	939	7.33	1.707	75.	Common Spindle, long Band, 16 Spin. on one side,
3.14	22. "	817	6.38	1.485	86.	" 16 Spin. across, 8 each side. [Av. ½ full.
3.14	22. "	750	5.86	1.364	94.	" 8 " " 4 " "
3.30	24. "	619	4.30	1.125	128.	Straight Spindle.
3.30	24. "	716	4.97	1.302	110.	" Same Frame, increased Speed.
3.30	24. "	978	6.79	1.779	81.	" " Further "
3.30	24. "	1,553	10.78	2.824	51.	" " " "
1.50	14. weft.	422	3.30	.768	167.	Taper Spindle.
1.50	14. "	627	4.90	1.140	112.	" " increased Speed.
1.50	14. "	1,053	8.23	1.915	67.	" " Further "
4.	30. warp.	853	6.64	1.551	83.	Bobbin one-half full.
4.	30. "	896	7.	1.629	78.5	" "
4.	30. "	1,006	7.86	1.830	70.	" "
4.	30. "	827	6.46	1.562	85.	" "
4.	30. "	776	6.06	1.410	90.5	" "
3.54	28. "	746	5.63	1.358	98.	" "
3.54	28. "	991	5.16	1.800	107.	" " Cochcco Dynamometer.
3.54	28. "	1,016	5.29	1.847	104.	" " two-thirds full, Amoskeag "
3.80	31. "	1,370	7.35	2.491	77.	Half full, damp day, next to wall of mill.
3.80	31. "	1,204	6.27	2.190	87.66	" Clear day, in centre of room.
3.80	31. "	736	5.75	1.337	95.75	" " "
3.80	31. "	910	7.11	1.655	77.	" " "
4.37	32. "	996	5.18	1.811	106.	Empty Bobbin, } Av'ge, 1,069.5 ft. lb.=1.945 H.P. Full " } =5.57 lb. Spin.=99. Spin. H P.
4.37	32. "	1,143	5.95	2.078	92.	
4.13	29. "	1,410	6.29	2.580	88.	Half-full Bobbin.
4.50	29. "	947	7.40	1.722	74.2	Frame not level, old Mill.
4.50	.. "	726	5.68	1.321	97.	Bands too tight and Bobbins too heavy.
4.50	.. "	726	5.68	1.321	97.	Bobbin lighter than last.
4.40	28.56 "	1,098	6.87	1.996	80.	Empty Bobbin, } Av'ge, 1,120 ft. lb.=2.037 H. P.= Full " } 7.01 lb. Spin.=78.5 Spin. H. P.
4.40	28.56 "	1,141	7.15	2.073	77.	
....	......	1,007	6.30	....	....	Empty Bobbin only on Spin. Rolls stopped.
....	......	1,050	6.56	....	....	Full " " "
4.40	28.56 "	1,253	7.83	2.378	70.	Empty " { Av'ge, 1,432 ft. lb. = 2.602 H. P. =
4.40	28.56 "	1,630	10.19	2.965	54.	Full " { 8.95 lb. per Spin.=61.5 Spin. H. P.
....	......	1,087	6.80	....	....	Empty Bobbin only on Spindle. Rolls stopped.
....	......	1,447	9.04	....	....	Full " " "
2.	15. "	1,395	7.92	2.536	70.	Empty " { Av'ge, 1,470 ft. lb. = 2.671 H. P. =
2.	15. "	1,575	8.77	2.807	62.	Full " { 8.35 lb. Spin.= 66. Spin. H. P.
0.95	9. "	1,270	6.61	2.309	83.	Two-thirds full Bobbin, tight Bolster.
0.95	9. "	886	4.61	1.610	119.	Half-full Bobbin, tight Bolster, reamed out.
4.	30. "	806	4.20	1.466	131.	" " Evidently full fast enough.

RING-SPINNING.—(*Continued.*)

Date.	Place.	Maker.	Diam Ring.	No. Spin.	Wt. per Spindle.	Rev. Spindle.	Rev. Roll.	Draft.
Jan., 1874	Mt. Vernon Mills,	Bridesburg Mfg.	1¾	204	8 oz.	5,050	128	7.
" "	Baltimore, Md.,	Co.,	"	204	"	5,050	128	7.
" "	"Carroll Spin.,"	"	"	204	"	5,050	80	7.
" "	"	"	"	204	"	5,050	80	7.
" "	"*	"	"	204	"	5,050	80	7.
" "	"	"	"	204	"	5,050	80	7.
" "	"	"	"	204	"	5,050	80	7.
" "	"	"	"	204	"	5,050	80	7.
" "	"	"	"	204	"	5,050	80	7.
Feb., 1874	Nashua Mfg. Co.,	Fales & Jenks,	"	128	"	5,930	94	7.
" "	{ Altered from Throstle,	"	"	128	"	5,930	94	8.66

SAWYER

Date.	Place.	Maker.	Diam Ring.	No. Spin.	Wt. per Spindle.	Rev. Spindle.	Rev. Roll.	Draft.
April, 1872	{ Appleton Mills, Lowell,	Lowell Ma. Shop,	"	160	4 oz.	6,050	120	8.
" "	"	Altered from old	"	128	"	3,055	66	8.
" "	"	Throstle-Frame,	"	128	"	4,027	88	8.
" "	"	"	"	128	"	5,000	107	8.
" "	"	"	"	128	"	2,950	83	8.07
" "	"	"	"	128	"	4,027	110	8.07
" "	"	"	"	128	"	4,027	110	8.07
" "	"	"	"	128	"	4,027	110	8.07
" "	"	"	"	128	"	4,115	117	8.07
" "	"	"	"	128	"	4,022	115	8.07
Oct., 1872	Stark Mills, Manchester, N. H.,	"	"	128	"	5,605	115	7.84
" "	"	"	"	128	"	5,069	104	7.84
" "	"	"	"	128	"	4,386	90	7.84
" "	"	"	"	128	"	3,753	77	7.84
Jan., 1873	"†	"	"	128	"	4,180	90	7.84
" "	"	"	"	128	"	4,900	104	7.84
" "	"	"	"	128	"	5,320	115	7.84
" "	"	"	"	128	"	6,320	135	7.84
" "	"‡	"	"	128	"	4,600	91	7.84
" "	"	"	"	128	"	5,230	106	7.84
" "	"	"	"	128	"	5,770	117	7.84
" "	"	"	"	128	"	6,350	130	7.84
Feb., 1873	{ Cocheco Mills, Dover, N. H.	Old Frame alter'd	1¼	128	3¾ oz.	6,002	78	7.90
Mar., "	Stark Mills, "	"	1¼	128	3¼ oz.	5,290	110	7.84
" "	King Philip, F. R.	Saco W. P. Co., New,	1⅜	160	"	6,260	65	7.31
" "	"	"	"	160	"	6,260	....	....
" "	Manville, R. I.,	Fales & Jenks, alt.,	1½	128	"	7,730	80	11.25
" "	"	"	"	128	"	7,730	80	11.25
April, "	{ Social Mill, Woonsock't, R.I.	Whitin, New,	1¼	192	"	6,448	77	7.52
" "	"	"	"	192	"	6,448	77	7.52

* Frame had been in operation three weeks longer. † Same frame. ‡ Similar.

RING-SPINNING.—(*Continued.*)

From	To	Ft. Lb. Frame.	Ft. Lb. Spind.	H. P. Frame.	Spindl's H. P.	Remarks.
0.85 hk	6. wp	1696.	8.31	3.080	66.	Mill cold, 60° Fahr. Oil at 65 cents per gallon.
0.85	6. "	1875.	9.19	3.409	60.	" damp " " 35 "
1.88	12. "	1661.	8.14	3.020	67.	" " " " 35 "
1.83	12. "	1500.39	7.35	2.727	75.	" " " " 65 "
1.83	12. "	1464.	7.18	2.662	77.	" " " " 65 "
1.83	12. "	1348.25	6.61	2.451	83.2	" warmer = 70°, 65 "
1.83	12. "	1134.	5.56	2.062	....	Ends down. Traveler stopped = 13.60 per cent.
1.83	12. "	1009.	4.95	1.835	....	Rolls stopped = 9.52 per cent.
1.83	12. "	884.	4.33	1.607	....	Bobbin off = 9.53 per cent. Spin. only 67.35 per ct.
2.50	22. "	1034.30	8.08	1.880	....	Empty Bobbin. Warm and clear.
2.50	22. "	1271.20	9.93	2.311	....	Full " "
		1152.54	9.	2.095	61.	Average " Bands heavy and hard.

SPINDLE.

From	To	Ft. Lb. Frame.	Ft. Lb. Spind.	H. P. Frame.	Spindl's H. P.	Remarks.
1.65	12.50 "	662.	4.14	1.204	133.	Average of 18 Tests. Bobbin half full.
1.65	12.50 "	207.	1.62	.376	340.	Bobbin half full. Loose Bands, 55 in., $\frac{1}{16}$ in diam.
1.65	12.50 "	326.	2.54	.593	216.	"
1.65	12.50 "	454.	3.55	.825	155.	"
1.65	13. wf.	211.	1.65	.384	333.	"
1.65	13. "	322.	2.52	.586	219.	"
1.65	13. "	343.	2.68	.623	205.	" Damp day, wet floor. Bands 54½ in.
1.65	13. "	303.	2.37	.551	232.	" Dry day. New Bands, 55½ in.
1.65	13. "	399.	3.12	.725	176.	" Common Ring Bands, ¼ diam. Dry.
1.65	13. "	417.	3.26	.758	167.	" " " " Damp.
1.61	12. wp	731.58	5.71	1.330	96.	Half-full Bobbin, 1.10 in. Band.
1.61	12.33 "	541.	4.26	.984	130.	" Speed reduced.
1.61	12.33 "	444.	3.47	.808	158.	" Speed further reduced.
1.61	12.33 "	361.	2.82	.656	180.	" " " "
1.61	12.33 "	451.	3.52	.820	156.	" 1-8 inch Band.
1.61	12.33 "	596.	4.66	1.083	118.	" Speed increased.
1.61	12.33 "	667.	5.21	1.218	105.	" Speed further increased.
1.61	12.33 "	890.	6.96	1.619	80.	" " " "
1.61	12.33 "	486.	3.80	.884	145.	" New Bolster, 1.10 in. loose Band.
1.61	12.33 "	611.	4.73	1.112	115.	" " Speed increased.
1.61	12.33 "	718.	5.61	1.300	95.5	" " Speed further increased.
1.61	12.33 "	899.	7.02	1.634	78.5	" " " " "
3.54	28. wp	577.	4.50	1.049	122.	" Spiral Bolster. [Rolls, 24 per cent.
1.61	12.33 "	547.	4.28	.995	129.	" Cyl. and Spin., 56 per ct., Traveler, 20 p. ct.,
6.	40. "	613.43	3.83	1.115	143.5	"
....	......	366.	2.28	.665	....	" = 60 per ct. of whole power for Cyl. & Spin.
7. dou.	40. "	493.	3.85	.897	143.	Empty Bobbin, 334 grains weight.
7. dou.	40. "	525.	4.10	.955	134.	Half-full Bobbin = 172 gr. yarn additional.
4.37	32. "	750.	3.91	1.364	140.	Empty " { Average, 846 ft. lb.= 1.538 H. P.
4.37	32. "	941.	4.90	1.711	112.	Full " { =4.40 ft. per Spin.=125. Spin. H.P.

RING-SPINNING.—(*Continued.*)

Date.	Place.	Maker.	Diam Ring.	No. Spin.	Wt. per Spindle.	Rev. Spindle.	Rev. Roll.	Draft
Jun. 2, 1873	Pacific Mills, Lawrence,	Whitin, altered to Saw. Spin.,	1⅝	160	3.75 oz	6,059	73	6.79
"	"	"	"	160	"	6,059	73	6.79
"	"	"	"	160	"	6,059	73	6.79
" 5, "	"	"	"	160	"	6,059	73	6.79
"	"	"	"	160	"	6,059	73	6.79
"	"	"	"	160	"	6,059	73	6.79
" 2, "	"	"	"	160	"	6,059	....	....
"	"	"	"	160	"	6,059	....	....
" 5, "	"	"	"	160	"	6,059	....	....
Sep. 30, "	Same Frame repeated.	"	"	160	"	6,059	73	6.79
"	"	"	"	160	"	6,059	73	6.79
"	"	"	"	160	"	6,059	73	6.79
Dec., "	Stark Mills, Manchester, N. H.,	Lowell Ma. Shop, Old Frame alter'd	1¾	128	3.75	5,820	97	9.48
"			"	128	"	5,820	97	9.48

RABBETH

Jan., "	Potonska Mills, N. Bedford, Mass.,	Fales & Jenks, Rabbeth Spind.	1½	480	3½ oz.	6,100	80	7.50
"	"	"	"	160	"	6,100	80	7.50
Mar., "	"	"	"	160	"	6,160	79	7.50
"	"	"	"	160	"	6,200	80	7.50
"	"	"	"	160	"	6,200	80	7.50
"	"	"	"	160	"	6,200	80	....
"	"	"	"	160	"	6,200	Spin	only
"	"	"	"	160	"	6,200	80	7.50
Jun. 2, "	Pacific Mills, Lawrence, Mass.,	Lowell Shop, al. to Rabbeth Spin. by Fales & Jenks,	1¾	160	3¾ oz.	6,059	73	6.79
"	"		"	160	"	6,059	73	6.79
"	"		"	160	"	6,059	73	6.79
"	"	"	"	160	"	6,059	73	....
"	"	"	"	160	"	6,059	73	....
" 5, "	"	"	"	160	"	6,059	73	6.79
"	"	"	"	160	"	6,059	73	6.79
"	"	"	"	160	"	6,059	73	6.79
"	"	"	"	160	"	6,059	....	....
Sep. 30, "	"	"	"	160	"	6,059	73	6.79
"	"	"	"	160	"	6,059	73	6.79
"	"	"	"	160	"	6,059	73	6.79

PEARL

Jun. 2, "	Pacific Mills,	Lowell Machine Shop, Pearl Spindle No. 3,	"	160	3⅞ oz.	6,059	73	6.79
"	"	"	"	160	"	6,059	73	6.79
"	"	"	"	160	"	6,059	73	6.79
"	"	"	"	160	"	6,059	....	....
"	"	"	"	160	"	6,059	....	....
" 5, "	"	"	"	160	"	6,059	73	6.79
"	"	"	"	160	"	6,059	73	6.79
"	"	"	"	160	"	6,059	73	6.79
"	"	"	"	160	"	6,059	....	....

SAWYER SPINDLE.—(Continued.)

From	To	Ft. Lbs. Frame.	Ft. Lbs. Spindle.	H. P. Frame.	Spindl's H. P.	REMARKS.
hanks. 4.40	28.50 wp	692.	4.32	1.258		Empty Bob., clear day, ⎰ Av. of 3 tests, 728. ft. lb.,
4.40	" "	724.	4.53	1.317	121	¼ full " " ⎱ 1.322 H. P., 4.55 lb. per
4.40	" "	766.	4.79	1.394		Full " " ⎱ Spin., 121 Spin. per H. P.
4.40	" "	726.5	4.54	1.321		Empty Bob., damp day ⎰ Av. of 3 tests, 753 ft. lb.,
4.40	" "	727.4	4.58	1.322	117	¼ full " " ⎱ =1.369 H. P.= 471 lb. per
4.40	" "	805.7	5.04	1.465		Full " " ⎱ Spin.=117. Spin. per H.P.
....	....	479.	3.	......	......	Empty Bob. only on Spin. Rolls stopped=69.5 pr.
....	....	554.	3.46	......	......	Full " " " [cent. of whole.
....	....	490.	3.06	......	......	Empty Bobbin only = 67.5 per cent.
4.40	" "	689.7	4.31	1.254		Empty Bobbin ⎰ Average, 758.6 ft. lb., 1.372 H. P.
4.40	" "	758.6	4.74	1.372	116	¼ full " " ⎱ 4.74 lb. per Spindle.
4.40	" "	827.6	5.17	1.534		Full " ⎱ 116 Spindles per H. P.
1.61	15.25 "	416.6	3.25	.757	169	Empty Bobbin, ⎰ Average, 470 ft. lb.= .854 H. P.=
1.61	" "	522.7	4.09	.950	135	Full " ⎱ 3.67 lb. per Spin.=150 Spin. H.P.

SPINDLE.

4.	30. warp	2285.	4.75	4.155	116 —	⎰ 3 Frames taken together average, ¼ full, ⎱ Full = 4.491 H. P. Empty = 3.977 H. P.
4.	" "	759.	4.73	1.380	116 +	Half-full Bobbin.
4.	" "	770.	4.81	1.401	114	" Centre of cylinder ¼ in. below whorl.
4.	" "	702.	4.39	1.277	125	" " " raised level with "
4.	....	584.	.....	......	......	Ends down, Traveler stopped = 16.93 per cent.
....	....	571.	.....	......	......	Roving Broken, Draught stopped = 1.70 per cent.
....	....	500.	3.125	=71p c	......	Rolls stopped = 10.17 per cent. [oiled.
4.	" "	690.5	4.315	1.255.	128	¼ full, Bands all put in order, and Frame well
4.40	28.50 "	609.	3.80	1.107	Av'ge. 132.5	Empty Bobbin, clear day.
4.40	" "	689.	4.31	1.253		¼ full Bobbin ⎰ Average, 664 ft. lb. = 1.208 H. P.=
4.40	" "	696.	4.35	1.265		Full " ⎱ 4.15 lb. per Spin.=132.5 Spin. H. P
....	....	460.	2.89	......	......	Empty Bobbin only = 75.5 per cent.
....	....	532.	3.32	......	......	Full " = 76.3 " [ster-step
4.40	" "	617.	3.85	1.122	Av'ge. 132	Empty Bob., damp day, but with fresh oil in Bol
....	" "	672.3	4.20	1.252		¼ full Bobbin ⎰ Average, 667.4 ft. lb. = 1.232 H. P. =
4.40	" "	712.8	4.45	1.290		Full " ⎱ 4.17 lb. Spin. = 132 Spin. H. P.
....	....	457.4	2.88	......	......	Empty Bobbin only on Spindle = 74 per cent.
4.40	" "	574.5	3.59	1.045	Av'ge, 141	" " Reduced in weight from 379 gr. to 214 gr
4.40	" "	627.7	3.92	1.141		¼ full Bobbin ⎰ Average, 694.8 ft. lb. = 1.136 H. P.=
4.40	" "	672.4	4.20	1.222		Full " ⎱ 3.90 lb. per Spin.= 141. Spin. H.P

SPINDLE.

4.40	" "	519.6	3.25	.945	Av'ge, 141	Empty Bobbin, clear day.
4.40	" "	664.3	4.15	1.208		¼ full Bobbin ⎰ Average, 623 ft. lb = 1.134 H. P.=
4.40	" "	687.5	4.30	1.250		Full " ⎱ 3.90 lb. Spin.= 141 Spin. H. P.
....	....	383.9	2.42	=74p c	......	Empty Bobbin on Spindle only. Rolls stopped.
....	....	526.8	3.29	=76p c	......	Full " " "
4.40	" "	578.9	3.62	1.053	Av'ge, 134	Empty Bobbin, damp, showery day.
4.40	" "	682.5	4.28	1.241		¼ full Bobbin ⎰ Average, 657.31 ft. lb.= 1.195 H.P.=
4.40	" "	710.5	4.44	1.292		Full " ⎱ 4.11 lb. per Spin.= 134 Spin. H. P.
4.40	....	430.	2.69	=74p c	......	Empty Bobbin only on Spin. Rolls stopped.

RING-SPINNING.—(*Continued.*)

Date.	Place.	Maker.	Diam Ring.	No. Spin.	Wt. of Spindle.	Rev. Spindle.	Rev. Roll.	Draft.
Sep. 30, '73	Pacific Mills,	Pearl Spin. No. 2,	1⅜ in	160	4½ oz.	6,059	73	6.79
"	"	"	"	160	"	6,059	73	6.79
"	"	"	"	160	"	6,059	73	6.79
" 25, "	Atlantic Mills,	"	"	176	"	5,719	80	7.60
"	"	"	"	176	"	5,719	80	7.60
"	"	"	"	176	"	5,719	80	7.60
" 27, "	"	"	"	176	"	5,719	80	7.50
"	"	"	"	176	"	5,719	80	7.60
"	"	"	"	176	"	5,719	80	7.60
" 25, "	"	"	"	176	"	5,285	103	7.60
"	"	"	"	176	"	5,285	103	7.60
"	"	"	"	176	"	5,285	103	7.60
"	"	"	"	176	"	5,448	90	7.60
"	"	"	"	176	"	5,448	90	7.60
"	"	"	"	176	"	5,448	90	7.60
" 26, "	"	"	"	176	"	5,448	66	7.60
"	"	"	"	176	"	5,986	72	7.60
" 27, "	"	"	"	176	"	5,735	70	7.60
"	"	"	"	176	"	5,735	70	7.60
"	"	"	"	176	"	5,735	70	7.60

BIRKENHEAD

Date	Place	Maker	Diam Ring	No. Spin	Wt. of Spindle	Rev. Spindle	Rev. Roll	Draft
Mar., 1873	Ric'd Borden Mill	Wm. Mason, Bir-	1⅜ in	192	4 oz.	5,100	69	8.14
"	Fall River, Mass.,	kenhead Spin.,	"	192	"	5,100	....	...
"	Amer. Linen Co.,	Higgins, altered	1½	160	"	5,880	69	7.09
"	"	to Birkenhead,	"	160	"	5,830	....	....
"	"	"	"	160	"	6,840	81	7.09
May, 1873	"	"	"	160	"	5,750	68	7.09
"	"	"	"	160	"	5,750	68	7.09
"	"	"	"	160	"	5,750	68	....
"	"	"	"	160	"	5,750	68	....
"	"	"	"	160	"	5,750	....	....

RICHARDSON

Date	Place	Maker	Diam Ring	No. Spin	Wt. of Spindle	Rev. Spindle	Rev. Roll	Draft
June, 1873	Ocean Mills, New-	Lowell Ma. Shop,	1⅜ in	208	....	5,200	67	7.04
"	buryport, Mass.,	Richardson &	"	208	...	5,440	71	7.04
"	"	Cumnock's Spin.	"	208	....	5,290	68	7.04
"	"	and Bolster,	"	208	....	5,000	65	7.04
"	"	"	"	208	....	5,290	68	7.04

EXCELSIOR

Date	Place	Maker	Diam Ring	No. Spin	Wt. of Spindle	Rev. Spindle	Rev. Roll	Draft
Oct., 1872	Frankford, Pa.,	Bridesburg Man-	1¼ in	204	4 oz.	2,960	62	6.
Dec., 1872	"	ufacturing Co.,	"	204	"	4,103	90	6.
"	Wingohocking M'l	"	"	204	"	5,081	111	6.
"	"	"	"	204	"	6,053	130	6.
"	"	"	"	204	"	6,923	148	6.
"	"	"	"	204	"	5,035	68	7.
"	"	"	"	204	"	6,020	79	7.
"	"	"	"	204	"	7,009	90	7.
"	"	"	"	204	"	8,026	102	7.

PEARL-SPINDLE.—(Continued.)

From	To	Ft. Lb. Frame.	Ft.Lb. Spin.	H. P. Frame.	Spindl's H. P.	Remarks.
4.40	warp. 28.50	732.14	4.58	1.331	Av'ge, 108.	Empty bobbin, Bolsters too tight fit. Damp day..
4.40	28.50	839.3	5.25	1.526		Half full Bob. ⎰ Av'ge, 815.48 ft. lb.=1.483 H. P. =
4.40	28.50	875.	5.47	1.591		Full " ⎱ 5.10 lb. per Spin.=108. Spin. H. P.
2.80	20.50	806.5	4.58	1.466	....	Empty Bobbin. Taken just as running in mill.
2.80	20.50	959.7	5.45	1.745	....	Full " [No cleaning.
2.80	20.50	883.	5.02	1.605	110.	Average. [and oiled.
2.80	20.50	685.5	3.90	1.246	....	Empty Bobbin. Same Frame, thoroughly cleaned
2.80	20.50	758.	4.31	1.378	....	Full "
2.80	20.50	721.8	4.10	1.312	134.	Average Bobbin.
1.87	14.	806.	4.58	1.465	....	Empty " Similar Frame, taken as running
1.87	14.	1000.	5.68	1.818	....	Full "
1.87	14.	903.	5.13	1.641	107.	Average " " "
2.	15.	824.60	4.68	1.500	....	Empty " Frame like last. Spindle half an
2.	15.	962.46	5.47	1.752	....	Full " [inch shorter.
2.	15.	893.53	5.07	1.621	110.	Average "
3.68	28.	701.75	3.986	1.273	138.	Half full " Same Frame, finer Yarn.
3.68	28.	789.47	4.486	1.435	122.	" " 4 P. M., same Frame, speed incr'd.
3.68	28.	694.9	3.94	1.263	....	Empty " 10 A. M., " cool morning.
3.68	28.	915.24	5.20	1.666	....	Full "
3.68	28.	805.08	4.57	1.465	120.	Average " " " "

SPINDLE.

From	To	Ft. Lb. Frame.	Ft.Lb. Spin.	H. P. Frame.	Spindl's H. P.	Remarks.
3.53	29.	817.71	4.25	1.487	129.	Half full " Bands very tight.
....	......	666.60	3.47	....	....	Spindle and Cylinder only = 81.50 per cent.
4.13	29.	630.66	3.94	1.147	139.	Bobbin half full.
....	......	516.	3.23	.938	....	Spindle and Cylinder only = 81 per cent.
4.13	29.	836.	5.22	1.520	105.	Bobbin half full. Too high speed for Frame.
4.13	29.	690.8	4.31	1.256	128.	" " Same Frame as above.
4.13	20.	680.	4.25	1.237	130.	" " " Bands eased.
....	......	593.	....	1.079	....	Ends down, Traveler stopped = 12.8 per cent.
....	......	553.	....	1.006	....	Weight off Rolls = 5.85 per cent.
....	......	540.	3.375	.982	....	Rolls stopped. Spin. and Cyl. only = 79.4 per ct.

SPINDLE.

From	To	Ft. Lb. Frame.	Ft.Lb. Spin.	H. P. Frame.	Spindl's H. P.	Remarks.
3.90	29.	975.	4.69	1.773	117.	Bob. ½ full. Banded in usual way. Belt too tight.
3.90	29.	1082.	5.20	1.967	106.	" Frame not level. Belt too tight.
3.90	29.	711.6	3.42	1.294	160.	" " Level. Belt easy.
3.90	29.	746.6	3.59	1.359	153.	" Belt tight. ⎰ Banded, 6 Spin. with long
3.90	29.	575.4	2.77	1.046	199.	" Belt easy. ⎱ Band across, 3 on each [side.

SPINDLE.

From	To	Ft. Lb. Frame.	Ft.Lb. Spin.	H. P. Frame.	Spindl's H. P.	Remarks.
2.	12.	326.	1.00	.593	344.	1st Frame tested of this Spindle.
2.	12.	565.	2.77	1.027	200.	2d " Bobbin half full in all cases.
2.	12.	769.	3.77	1.398	146.	" "
2.	12.	1038.	5.09	1.888	108.	" "
2.	12.	1273.	6.24	2.314	88.	" "
2.90	20.5	620.	3.04	1.127	181.	" "
2.90	20.5	804.	3.94	1.462	140.	" "
2.90	20.5	1009.	4.94	1.834	111.	" "
2.90	20.5	1206.	6.11	2.266	90.	" "

RING-SPINNING.—(Continued.)

Date.	Place.	Maker.	Diam. Ring.	No. Spin.	Wt. Spin.	Rev. Spindle.	Rev. Roll.	Draft
Dec., 1872	Frankford, Pa.,	Bridesburg Mfg. Co.,	1¼ in	204	4 oz.	9,067	118	7.
"	"	"	"	204	"	5,081	62	7.50
"	"	"	"	204	"	6,053	72	7.50
"	"	"	"	204	"	7,009	84	7.50
"	"	"	"	204	"	8,026	96	7.50
"	"	"	"	204	"	9,067	108	7.50
"	"	"	"	204	"	10,071	121	7.50
Mar., 1873	Stark Mills, Manchester, N. H.	"	1 7/16	204	"	4,900	116	8.30
"	"	"	"	204	"	5,050	118	8.30
"	"	"	"	204	"	5,200	120	8.30
"	"	"	"	204	"	4,736	120	8.30
"	"	"	"	204	"	5,264	131	8.30
June, 1873	"	"	"	204	"	4,287	110	8.30
April 2, '73	Mannville, R. I.,	"	1½	204	"	8,300	80	5.71
"	"	Cast-Iron Step and Bolster,	"	204	"	8,300	80	....
"	"	"	"	204	"	8,300	....	....
"	"	"	"	204	"	8,300	....	....
"	"	"	"	204	"	8,300	....	....
"	"	"	"	204	"	......	....	....
April 3, '73	"	"	"	204	"	9,300	91	5.71
"	"	"	"	204	"	8,300	80	5.71
"	"	"	"	204	"	7,550	73	5.71
"	"	"	"	204	"	6,870	67	5.71
"	"	"	"	204	"	6,390	62	5.71
April 7, '73	"	Brass Step and Bolster,	"	204	"	9,600	100	5.71
"	"	"	"	204	"	8,740	91	5.71
"	"	"	"	204	"	8,300	86	5.71
"	"	"	"	204	"	7,650	80	5.71
"	"	"	"	204	"	6,870	72	5.71
"	"	"	"	204	"	6,150	64	5.71
April 9, '73	"	Ring changed,	1 7/16	204	"	9,440	98*	5.71
"	"	"	"	204	"	8,300	86	5.71
"	"	"	"	204	"	7,450	77*	5.71
"	"	"	"	204	"	6,870	70*	5.71
"	"	"	"	204	"	6,150	64	5.71
Apr. 11, '73	"	"	"	204	"	10,040	104	5.75
"	"	"	"	204	"	9,080	94	5.75
"	"	"	"	204	"	8,300	86	5.75
"	"	"	"	204	"	7,650	80	5.75
"	"	"	"	204	"	6,970	73	5.75
"	"	"	"	204	"	6,000	63	5.75
Nov., 1873	Weetamoe Mills, Fall River, Mass.,	"	1½	204	"	5,760	90	9.06
"	"	"	"	204	"	5,760	90	9.06
Jan., 1874	Clipper Mills, Baltimore, Md.,	"	1¼	132	"	4,660	122	6.50
"	"	"	"	132	"	4,660	122	6.50
"	"	"	"	132	"	2,700	70	6.50

EXCELSIOR SPINDLE.—(*Continued.*)

From	To	Ft. Lb. Frame.	Ft. Lb. Spin.	H. P. Frame.	Spindl's H. P.	REMARKS.
2.00	20.5 wp.	1508.	7.89	2.742	74.5	2d Frame.
4.	30.5 "	546.	2.67	.993	205.	" "
4.	30.5 "	705.	3.45	1.282	159.	" "
4.	30.5 "	929.	4.55	1.649	121.	
4.	30.5 "	1174.	5.75	2.134	95.5	
4.	30.5 "	1475.	7.23	2.862	76.	
4.	30.5 "	1754.	8.60	3.190	64.	
1.50	13. weft.	864.	4.23	1.571	130.	Bobbin half full. Bands too tight.
1.50	13. "	896.	4.39	1.629	125.	{ Cyl. & Spin. 68.4 p. c., Twist 16 do., Rolls 15.6 do.
1.50	13. "	963.	4.72	1.753	116.5	{ Bobbin half full. New Bands, larger,
1.50	13. "	693.	3.40	1.260	162.	{ Cyl. & Spin. 61.4 p. c., Twist 20 do., Rl's. 18.6 do.
1.50	13. "	772.	3.78	1.404	145.	{ Bobbin half full. New Bands, larger,
1.50	13. "	571.	2.80	1.038	196.5	{ C. & S. 66.66 p. c., Twist 18.02 do., Rolls 15.32 do. Bobbin half full.
7.	40. warp	993.33	4.87	1.806	113.	Half full. New Frame, a new mill. Cold and wet.
....	....	893.33	....	1.624	....	" Yarn broken, Trav'r stop'd = — 10.07. p. c.
....	....	843.33	....	1.530	....	Roving " Draught " = — 5.03 "
....	....	816.66	....	1.485	....	Top Rolls off = — 2.69 "
....	....	766.66	3.76	1.394	....	Bottom Rolls stopped = — 5.08 "
....	....	100.	....	.182	....	{ Bands off, Spindles stopped = — 67.11 " { Cylinder only = — 10.07 "
7.	40. "	1233.33	6.04	2.242	91.	Half full. Weather warmer, but mill damp.
7.	40. "	916.86	4.50	1.665	122.	" and cool all through the experiments.
7.	40. "	772.73	3.79	1.405	145.	" " " " "
7.	40. "	700.	3.46	1.273	159.	" " " " "
7.	40. "	646.	3.17	1.175	174.	" " " " "
7.	40. "	1372.	6.73	2.495	82.	" Twist-gear changed. Colder than last test.
7.	40. "	1140.	5.59	2.073	98.	" " " "
7.	40. "	983.33	4.84	1.788	114.	" " " "
7.	40. "	833.33	4.08	1.576	135.	" " " "
7.	40. "	755.	3.70	1.373	149.	" " " "
7.	40. "	714.	3.50	1.298	157.	" " " "
7.	40. "	1190.	5.83	2.163	94.2	" Mill still very cold and damp.
7.	40. "	953.	4.67	1.733	118.	" " " "
7.	40. "	834.	4.09	1.517	134.5	" " " "
7.	40. "	733.33	3.59	1.333	153.	" " " "
7.	40. "	573.5	2.81	1.053	195.	" " " "
7.	40. "	1283.33	6.28	2.333	87.5	" Warm day. Wind S. W.
7.	40. "	983.6	4.82	1.788	114.	" " "
7.	40. "	833.33	4.08	1.515	135.	" " "
7.	40. "	743.33	3.64	1.351	151.	" " "
7.	40. "	643.33	3.15	1.190	175.	" " "
7.	40. "	500.	2.45	.909	224.	" " "
4.	36. weft.	640.	3.14	1.165	175.	Empty Bobbin.
4.	36. "	656.25	3.21	1.193	171.	Full "
1.	6.50 wp	607.43	4.60	1.104	} Av. } 111.	Empty Bobbin } Average, 4.95 lb. per Spindle.
1.	6.50 "	700.33	5.21	1.274		Full "
1.	6.50 "	319.	2.42	.580	228.	" "

RING-SPINNING.—(Continued.)

Date.	Place.	Maker.	Diam. Ring.	No. Spin.	Wt. Spin.	Rev. Spindle.	Rev. Roll.	Draft
June, 1872	Amosk'g Mills,	Amoskeag Co.,	1¾ in	128	....	5,154	61	7.68
"	"	"	"	128	....	6,187	68	7.68
"	"	"	"	128	....	3,506	74	9.35
Oct., 1872	"	"	"	128	...	3,090	66	9.35
"	"	"	"	128	....	4,050	82	9.35
"	"	"	"	128	....	5,672	115	9.35
"	"	"	"	144	....	4,600	70	7.26
"	"	"	"	144	....	5,250	77	7.26
"	"	"	"	144	....	6,127	90	7.26
"	"	"	"	144	....	7,355	107	7.26
Feb., 1874	"	"	"	144	....	5,004	66	7.26
"	"	"	"	144	....	5,700	75	7.26
"	"	"	"	144	....	6,316	83	7.26
"	"	"	"	144	....	7,384	97	7.26
"	"	"	"	144	....	8,380	110	7.26

PUSEY SPINDLE.

Jan., 1874	Clipper Mill,	Pusey Bros., Wil-	2¼ in	132	....	3,132	108	4.
"	Baltimore, Md.,	mington, Del.,	"	132	....	3,132	108	4.
"	Wilmington, Del.,	"	1¾ in	132	....	6,026	84	8.86
"	"	"	"	132	....	6,026	84	8.86
"	"	"	"	132	....	7,030	98	8.86
"	"	"	"	132	....	7,030	98	8.86

PERRY SPINDLE (*Dead*).

From	To	Ft. Lb. Frame.	Ft. Lb. Spin.	H. P. Frame.	Spindl's H. P.	REMARKS.
4.42	34 warp	493.	3.85	.896	143.	Bobbin half full. Set on Whorl which revolves around Spindle, the latter free to move.
4.42	34 "	608.	4.75	1.106	116.	Bobbin half full.
1.50	14 weft.	339.	2.65	.616	208.	" "
1.50	14 "	311.	2.43	.565	226.	" "
1.50	14 "	440.	3.44	.801	160.	" "
1.50	14 "	669.	5.23	1.217	105.	" "
3.30	24 warp	439.	3.05	.800	180.	" "
3.30	24 "	518.	3.60	.941	153.	" "
3.30	24 "	716.	4.97	1.202	110.	" "
3.30	24 "	956.	6.64	1.738	83.	" "
3.30	24 "	437.5	3.04	.795	181.	" "
3.30	24 "	514.	3.57	.935	155.	" "
3.30	24 "	639.	4.44	1.162	124.	" "
3.30	24 "	804.	5.58	1.462	98.	" "
3.30	24 "	1000.	7.	1.818	78.5	" "

(*Dead.*)

0.75	3 warp	1145.84	8.08	2.083	Av. 67.	Empty Bobbin. ▲		
0.75	3 "	1218.75	9.23	2.216		Full "	= 8 oz.	Yarn.
2.25	20 "	404.35	3.06	.735	Av. 156.25	Empty "		
2.25	20 "	525.22	3.98	.955		Full "	= 2 oz.	"
2.25	20 "	521.21	3.96	.949	Av. 123.5	Empty "		
2.25	20 "	712.80	5.40	1.278		Full "	= 2 oz.	"

▲ Bobbin placed on a tube which revolves around Spindle.

MULE-SPINNING.

Date.	Place.	Description.	No. Spin.	Ac. Rev. Spindle.	St'ch.	Sec. St'ch.
July, 1871	Amosk'g Mill, N.H.	1 pair Smith Mules,	896	4,500	65 in	19¼
"	Langdon, "	1 Saco W. P. Co. Improved Sharpe & Roberts	704	5,000	60	19.
Aug., 1871	Manchaug, Mass.,	1 Saco W. P. Co. Curtis, Parr & Madely.	696	5,000	64	22.
"	" "	" "	552	5,000	64	22.
Sept., 1871	Wauregan, Conn.,	1 Curtis, Parr & Madely,	696	4,860	62	21.66
"	" "	1 Marvel & Davol, Sharpe & Roberts,	516	4,560	61	21.
Aug., 1871	Manchaug, Mass.,	1 Wm. Mason & Co., old,	832	4,170	60	19.33
Sept., 1871	Durfee Mills,	1 Wm. Mason & Co., Warp $\frac{5}{16}$ Gau., new.	768	4,600	60	18.
"	Fall River, Mass.,	1 Wm. Mason & Co., 1$\frac{5}{16}$ Gauge, 5 months old, 1$\frac{5}{16}$ oz. warp,	768	4,600	60	18.
Jan., 1873	Whittenton, Taunton, Mass.,	1 Wm. Mason, " " "	576	3,000	60	18.
Mar., 1873	Haydensville, Mass.	1 " 1¼ Gauge,	468	4,200	60	17.5
Apr., 1873	Salmon Falls, N. H.	1 Saco W. P. Co., S. & R.,	560	3,850	60	17.
"	Appleton Mills, Lowell,	1 Platt Bros. 1⅜ Gauge,	600	2,460	64	15.5
"	"	Same Mule, Twist ch'd,	600	3,220	64	17.5
May, 1873	Rockport, Mass.,	1 Franklin Foundery, Geared Mule,	600	3,470	60	23.33
"	" "	1 Saco W. P. Co., F. F. Pattern,	544	3,850	60	18.
"	Masconomet Mill,	1 Wm. Mason, 1¼ Ga., old,	480	3,690	60	19.5
"	Newburyp't, Mass.,	" " "	572	3,690	60	19.5
Nov., 1873	Essex, Paterson, N. J.,	1 Curtis, Parr & Co., 1⅜ inch Gauge.	704	3,700	63	22.
"	"	Marvel, Davol & Co., 1¼ inch Gauge,	600	3,550	60	18.
Mar., 1873	Granite Mill, F. R.,	1 Platt Bros. & Co. 1⅜ ga. warp.	552	4,713	64	19.
Nov., 1873	" " "	Same Mule,	552	5,300	64	17.
Mar., 1873	Slade " "	1 Mason, 1$\frac{5}{16}$ Gau. warp,	556	4,480	60	18.
May, 1873	Weetamoe, "	1 Parr & Curtis 1$\frac{5}{16}$ ga. wp.	544	5,106	63	17.
"	" " "	1 " 1$\frac{3}{16}$ " wef.	600	4,113	63	17.
Nov., 1873	Westville, Taunton,	1 Wm. Mason 1¼ " "	480	3,100	60	18.

MULE-SPINNING.

Draft.	From	To	START.		DRAFT & TWIST		BACKING.		Average Ft. Lb.	Lb. per Spindle.	Average H. P.	Spindles per H.P.
			Sec.	Ft. Lb.	Sec.	Ft. Lb.	Sec.	Ft Lb				
8.09	3.33	30. weft.	*	*	*	*	*	*	1,472	1.64	2.677	335.
7.17	4.76	33. "	2.	1,604	12	1,185	5	349	1,146	1.63	2.084	338.
8.74	5.95	52. "	4.	2,912	14	1,912	4	912	1,912	2.75	3.476	200.) A
8.74	5.95	52. "	4.	2,772	14	1,772	4	772	1,772	3.21	3.222	173.) A
8.71	7.	58. "	4.	2,701	13⁶⁶	1,701	4	701	1,701	2.59	3.093	225.
9.15	4.33	40. "	3.	2,220	13	1,660	5	600	1,660	3.22	3.019	171. B
8.40	4.	30. "	3.	2,579	13	1,592	5	605	1,592	1.91	2.895	287.
7.70	3.85	29. warp.	2.	2,918	{9 / 2}	{1,839 / 1,153}	5	388	1,486	1.93	2.702	284. } †
7.70	3.85	29. "	2.	2,892	{9 / 2}	{1,842 / 1,114}	5	383	1,473	1.92	2.078	287. }
8.50	2.	17. weft.	....	.....	12	1,050	6	478	859	1.46	1.562	370.
...	....	27. "	....	.....	11⁶	1,235	6	374	941	2.00	1.710	274.
8.56	3.62	31. "	2.	1,536	{10 / 1}	{1,082 / 627}	4	400	955	1.70	1.736	322.
8.	1.65	18. "	2.5	2,512	9	2,078	4	813	1,700	2.83	3.091	194.
7.95	1.65	12.50 wp	2.5	2,946	{9 / 2}	{2.503 / 1,742}	4	813	2,031	3.38	3.693	163.
7.	3.80	26. weft.	4.33	1,647	14	1,314	5	368	1,173	1.96	2.133	281.
7.21	3.80	32. "	3.	1,694	10	1,254	5	444	1,102	2.03	2.004	272.
8.25	4.50	36. "	2.	1,560	12⁶	1,357	5	372	1,124	2.34	2.045	235. } c
8.25	4.50	36. "	2.	1,613	12⁶	1,405	5	421	1,174	2.29	2.135	240. }
7.30	4.	30. warp.	2.	2.244	15	1,625	5	475	1,404	2.	2.533	275·
8.10	3.33	30. weft.	1.	1,207	12	873	5	367	734	1.225	1.335	450.
7.47	3.88	29. warp.	3.	3,104	11	2,195	5	695	1,944	3.525	3.535	156. D
7.47	3.88	29. "	3.	3,127	9	2,217	5	854	1,977	3.58	3.594	153.2
7.50	3.80	28. "	3.	1,967	8	1,517	7	363	1,099	1.976	1.998	279.
7.15	4.15	29.5 "	2.	2,620	10	2,046	5	517	1,664	3.06	3.025	180.
8.43	4.15	35. weft.	2.	2,098	10	1,512	5	465	1,257	2.10	2.287	262.
10.77	1.95	21. "	2.	1,009	10	855	6	309	690	1.46	1.255	382.5

* Mules balanced each other. † Companion Mules.
A Mules rather new and stiff. B Mules old and not quite level.
C Mules quite old. D Belt new and slipped. Not up to proper speed.

COTTON-LOOMS.

Date.	Place.	Description.	Width. (Inches.)	No. Warp	No. Weft.	Picks Clo'h per inch.	Picks p. min.
April, 1871	Stark Mills,	Lowell M. S., Sheeting	36	16.	16	56 × 48	128
June, 1871	Amoskeag Mills,	Amosk'g Co., Ticking,	28	10.	....	48 × 50	118
"	"	" Print Cloth,	28½	30.	33	64 × 64	120
July, 1871	"	" Fine Sheeting,	40	24.	30	64 × 80	125
"	"	" "	49	24.	30	"	120
"	"	" "	60	24.	30	"	116
"	"	" "	100	24.	30	"	70
"	"	" Fancy Gingham	28	13.5	16	56 × 56	82
"	"	" " [Goods,	28	24.	28	"	94
"	"	" simple Check"	28	24.	28	"	94
Aug., 1871	Manchaug,	Saco W.P.Co.Plain "	37	46.	52	....	140
"	"	" " "	37	46.	52	....	138
"	"	" " "	37	46.	52	....	154
"	"	Whitin's " "	36	46.	52	....	151
"	"	" " "	28	30.	32	64 × 64	150
"	"	" " "	28	30.	32	"	130
1856	Ind. Orchard,	Lowell M. S. " "	36	23.	24	64 × 72	126
"	"	" "	39	23.	24	64 × 68	126
Feb., 1872	Langdon Mills,	Whitin's " "	40	30.	33	88 × 100	120
"	"	" " "	40	33.	38	72 × 80	120
Mar., 1872	Haydensville,	" " "	36	23.	27	64 × 64	120
April, 1872	Salmon Falls,	Saco W.P.Co. " "	36	25.	31	64 × 72	125
Jan., 1872	Whittenton Mills,	Mason's Drop-box L'm	36	9.	11	"	112
"	"	1 Thomas's Patent "	36	9.	11	"	118
"	"	Mason's 4 Harness "	36	9.	11	"	115
"	"	" " "	36	9.	11	"	125
"	"	1 Revolving Box,	36	9.	11	"	110
"	"	Mason's 4 Harness,	36	9.	11	"	120
"	"	1 Crompton, New Pat.	36	9.	11	"	118
"	"	" Old "	36	22.	24	"	118
May, 1872	Rockport Mills,	W. Mason's Drill, "	33	28.	32	83 × 68	145
"	"	" " "	36	28.	32	"	140
"	"	" " "	40	28.	32	"	135
"	"	" " [Pat.	40	28.	32	"	140
"	"	Lewiston Sh., Mason's	40	28.	32	"	140
"	"	Wm. Mason, "	40	28.	32	"	117
"	Masconomet "	" Plain Sheeting,	40	30.	36	80 × 84	130
"	"	Lesley, "	40	30.	36	"	130
"	"	" "	48	30.	36	"	122
"	"	" "	36	30.	36	"	130
Nov., 1872	Essex, Pat'n, N.J.	Mosquito-Net Loom,	72	30.	30	12 × 12	88
"	Dale Mfg. Co., "	Jacquard on Silk Serge	27	....	....	....	100
"	"	Fancy Braid Loom, } Doub. Bk., 48 Shut. }	96	....	....	....	68
April, 1873	Clinton, Woon'kt	Whitin's Plain Loom,	40	31.	32	64 × 72	129
"	Social, "	" "	40	31.	32 {	5.06 × 68 7.76 × 84 }	139
Sept., 1873	Atlantic, Law'nce	Lowell Mach. Shop,	40	22.	24	64 × 68	145
"	"	" "	36	15.	16	52 × 52	156
"	"	" "	36	20.	20	56 × 56	156
Nov., 1873	Weetamoe, F. R.,	Kilburn, Lincoln &Co.	28	29.	35	64 × 64	154
"	Granite, "	Davol & Co.,	28	29.	35	"	154

COTTON-LOOMS.

Ft. Lb. per Sec. Loom.	H. Power Loom.	No. Tes'd	Total H. P.	Looms per H. P.	Remarks.
75.23	0.160	8	1.282	6.25	Dynamometer driving Counter Shaft.
106.15	.193	20	3.855	5.18	" " " "
35.14	.064	20	1.278	15.62	" " " "
51.15	.093	12	1.117	10.75	" " " "
62.15	.113	9	1.019	9.	" " " "
64.00	.118	15	1.779	8.47	" " " "
89.10	.162	3	.486	6.17	" " " "
96.82	.176	10	1.760	5.68	" " " "
86.35	.157	10	1.568	6.37	" " " "
48.95	.089	15	1.333	11.23	" " " "
65.	.118	1	.118	8.47 ⎫	" " " " ⎧ New Looms
72.05	.131	1	.131	7.63 ⎬	" " " " ⎨ just started.
86.35	.157	1	.157	6.37 ⎭	" " " " ⎩
85.80	.156	4	.624	6.41 ⎫	" " " " ⎧ Old Looms.
64.35	.117	4	.469	8.55 ⎬	" " " " ⎨
53.90	.098	4	.391	10.20 ⎭	⎩ " "
60.	.109	10	1.090	9.17	" " " "
64.50	.117	10	1.170	8.55	" " " "
115.7	.210	8	1.682	4.76	" " " "
74.4	.135	8	1.082	7.41	" " " "
71.86	.131	6	.786	7.83	" " " "
59.60	.108	10	1.084	9.23	" " " "
166.	.302	1	.302	3.31	Dynamometer applied direct to Loom. Doubtful
96.43	.175	1	.175	5.71	" " " "
102.5	.183	1	.183	5.47	" " old Loom, "
85.70	.156	1	.156	6.41	" " new " "
80.	.145	2	.290	6.90	" " C.-shaft. No doubt correct.
87.16	.158	1	.158	6.33	" " Loom quite new. Doubtful.
141.17	.257	1	.237	3.89	" " " "
116.2	.212	1	.212	4.72	" " " old, "
63.44	.116	4	.464	8.62	. " Counter-Shaft. Correct.
77.60	.144	2	.288	7.09	" " " New Loom, "
55.	.100	1	.100	10.	" " " Old " "
63.33	.115	3	.345	8.70	" " "
64.44	.117	3	.357	8.55	" " "
50.86	.092	4	.370	10.87	" " "
73.	.133	2	.266	7.52	" " "
80.	.145	2	.289	6.90	" " " Old Loom.
76.	.138	1	.138	7.25	" " " "
57.	.104	2	.207	9.61	" " " "
50.	.091	8	.727	11.	" " " "
32.5	.059	1	.059	17.	" " direct to Looms.
44.27	.080	1	.080	12.5	" " "
61.12	.111	14	1.561	9.	" " to Shaft.
93.62	.170	12	2.043	6.	" · "
92.6	.168	4	.673	6.	" " "
108.	.196	12	2.351	5.10	" " "
108.	.196	12	2.351	5.10	" " "
63.44	.115	10	1.153	9.	" " "
56.50	.103	12	1.233	10.	" " "

COTTON-SPOOLERS.

Date.	Place.	Description.	No. Spin.	Rev. Spin.	No. Yarn.	Lb. p. Spin.	H. P.
May, 1857	Ind. Orc'd, Mass.	Lowell Ma. S'p, upright,	100	936	23	1.18	0.215
June, 1871	Amoskeag, No. 3,	Amoskeag Shop,	80	700	10	2.35	.342
Aug., "	Manchaug,	Saco W. P. Co.,	100	663	46	.92	.167
Jan., 1872	Whittenton,	Mason, "Skein Sp'ler,"	60	....	16	3.11	.340
Mar., 1873	Haydensville,	Whitin, upright, "	80	800	30	3.62	.527
April, "	Salmon Falls,	Saco W. P. Co. "	100	700	25	1.80	.327
May, "	Rockport,	" "	70	1,000	28	2.82	.359
" "	Masconomet,	Lewiston Ma. Shop,	80	600	30	1.57	.228
" "	Weetamoc,	William Mason,	130	640	29	.81	.192
Sept., "	Atlantic Mill,	Lowell Ma. Shop,	96	700	15	1.62	.282
Nov., "	Granite Mill,	George Draper & Son,	80	630	29	1.28	.186

COTTON-TWISTERS.

Date.	Place.	Description.	Diam Ring.	No. Spin.	Rev. Spin.	Lb. p. Spin.	H. P.
June, 1871	Derry Mills,	{ Ring Frame used as } { Cass. Yarn-Twister, }	1¼ in	100	5,186	12.50	2.273
Nov., "	Stark Mills,	Duck "	3¼	80	2,812	10.	1.467
Jan., 1872	Whittenton Mill,	Mason's Ring,No. 17 Y'n	2¼	48	3,400	9.23	.805
Nov., "	Dale Silk Mill,	English Flier-Twister,	....	100	3,897	4.30	.783
" "	Paterson, N. J.,	" "	....	112	3,005	3.30	.673

COTTON-WARPERS.

Date.	Place.	Description.	No. Yarn.	No. Ends.	H. P.
June, 1871	Amoskeag,	....	10	230	0.171
Aug., "	Manchaug,	Lewiston English Box,	46	383	0.118
Jan., 1872	Whittenton,	Mason,	9	240	0.119
May, "	Masconomet,	Lewiston,	32	400	0.177
Nov., "	Granite,	English,	29	358	0.113

COTTON-DRESSERS.

Date.	Place.	Description.	No. Yarn.	No. Ends.	Yards p. Min	Lb. p. Sec.	H. P.
June, 1871	Amoskeag,	Amoskeag Co., Old Style	10	1,872	4	627	1.141
Jan., 1872	Whittenton,	Mason's "	9	1,920	10	1,177	2.139
Mar., "	Haydensville,	Whitin's "	23	2,300	7	1,060	1.927

SLASHERS.

May, 1871	Amoskeag,	Howard & Bullough,	28	2,720	20.	869.	1.581
Aug., 1872	Manchaug,	" "	46	2,298	23.57	855.	1.555
May, "	Rockport,	" "	28	2,592	24.	702.	1.277
April, 1873	Clint'n, Woon'kt,	" "	31	2,800	38.57	583.[33]	1.061
May, "	Weetamoe, F. R.,	" "	29	1,728	30.	395.	.702

MISCELLANEOUS MACHINERY AND TOOLS.

Date.	Place.	Description.	Revolu.	Ft. Lb.	H. P.
June, 1871	Amoskeag Mills,	1 Filling-Winder, on No. 34 Yarn, 100 Spindles,	2,910	793.	1.442
Jan., 1872	Whittenton "	1 Fil. Wind., 17 Yarn, 80 Spin.,	2,000	484.	.870
" "	" "	1 Reel; Skeins per 60 Spin.,	180	76.5	.140
" "	" "	1 Folding Ma., 70 yds. per min.	....	164.	.300
July, 1871	Langdon "	1 " " 75 "	....	69.	.126
May, 1872	Rockport,	1 " " 70 "	....	216.	.393
July, 1871	Langdon Mills,	1 Cloth Shear, 4 Blades,	2,000	1171.	2.130
		Fan to same,	516	526.	.957
		Total,	......	1697.	3.087
Jan., 1872	Whittenton,	1 Cloth Shear, 5 Blades & Fan,	2,070	2275.	4.136
May, "	Rockport, Mass.,	1 Cloth Shear, 3 Blades and Fan, 1,200 Rev. [hard wood	2,056	1645.	3.000
Mar., "	" "	1 Cir. Saw, 18 in. diam., 3 in.	1,300	700.	1.273
Nov., "	Paterson, N. J.,	1 " 9 in., 1 in. pine wood,	4,000	900.	1.637
" "	Whitney's Sewing Ma. Man.,	1 Small engine Lathe, on 3-8 inch Iron, [cut, 1 in. Iron,	....	51.	.092
May, "	Rockport, Mass.,	1 6 ft. Windsor Lathe, heavy	....	116.	.212
" "	" "	1 Upright Drill, 3-4 in. Drill,	....	88.	.160
Nov., "	Paterson, N. J.,	1 Upright Press Drill, 4 Sp., only 1 cutting, 1-4 in. hole,	....	168.75	.307
" "	" "	1 Crank-Planer, 2 in. stroke,	....	125.	.227
May, "	Rockport, Mass.,	1 5 ft. " 4 ft. "	....	135.	.245
Nov., "	Paterson, N. J.,	1 Profiling Machine, 1-4 in. Cutter, quick speed, [speed,	....	152.	.276
" "	" "	1 Pro. Ma., 1½ in. Cutter, slow	....	147.	.267
" "	" "	1 Milling Ma., small work,	....	121.	.220
" "	Whitney's Shop, Paterson, N. J.,	1 Small screw-cutting Engine, ¼ in. Screws,	....	122.	.222
" "	" "	3 Polishing-Wheels, 12 in. Diam., 1½ in. Face,	1,000	633.	1.151
May, 1873	Manch'r P. W.,	1 Log-Wash. Ma. in Bleach'y,	....	....	....
		1 Roll, 10ft. long, 20 in. Diam.	100	7650.	13.818
		1 " " 17 "	120		
" "	Underhill Edge-Tool Co., Nashua, N. H.,	1 Grindstone, 6 ft. Diam., 12 in. face, grinding Axes,	84	1680.	3.055
June, "	Collins Axe Co., Collinsville, Ct.,	1 Grindstone, 6.6 in. Diam., 13 in. face, grinding Axes, in Wood Boxes,	175	6260.	11.383
" "	" "	do. do. in Iron "	175	5263.	9.57
" "	" "	Stone in Revolution alone,	175	2807.	5.103
" "	" "	1 Stone, 3 ft. 10 in. Diam., 11 in. Face, grinding in Wood Boxes.	229	4300.	7.810
" "	" "	1 do., 2 ft. 8 in. Diam., 12½ in. Face, grind. in Wood Boxes,	229	3645.	6.627
" "	" "	1 Polishing Wh'l, 1 ft. Diam., 3 in. face, on Axes.	1,320	658.	1.200

FLAX MACHINERY.

Date.	Place.	Description.	Size.
April, 1871	Stark Mills, Manchester, N. H.,	1 1st Drawing-Frame, "Tow,"	4 Deliveries,
" "	"	1 2d " "	6 "
" "	"	1 Roving Frame, "	48 Spindles,
" "	"	1 Fairbairn's Spinning Fra. "	108 "
" "	"	1 Long Line Spreader, "Flax,"	1 Delivery,
" "	"	1 " 1st Drawing, "	2 Deliveries,
" "	"	1 " 2d " "	6 "
" "	"	1 " Roving Frame, "	48 Spindles,
" "	"	1 Fairbairn's Wet Spinning-Frame	116 "
" "	"	1 Lawson's " " " "	116 "
Nov., 1872	Arkwright Mills, Paterson, N. J.,	1 Long Line Spreader,	1 Delivery,
" "	"	1 2d Drawing-Frame,	3 Deliveries,
" "	"	1 Roving-Frame,	40 Spindles,
" "	"	1 Wet Spinning,	96 "
" "	"	1 Twine-Polisher,	Ordi'ry Pressure,
" "	"	"	Ex. Heavy "

WOOL MACHINERY.

WOOL-CARDS.

Date.	Place.	Description.	Width.	Diam.	Revolu.	Ft. Lb.	H. P.
Jan., 1871	Derry Mills,	{ 1 Davis & Furber 2d Breaker, 6 Workers.	40	48 in.	96	500	.910
Oct., 1871	Manchester P.W.	{ 1 Double Cylin. Card, 10 Work.,	40	48 "	130	700	1.273
" "	"	"	40	48 "	82	659	1.179

WOOL-JACKS.

Date.	Place.	Description.	Revolu.	DRAFT. Ft. Lb.	DRAFT. H. P.	TWIST. Ft. Lb.	TWIST. H. P.
June, 1871	Derry Mills,	{ 1 Davis & Furber, 200 Spin.,	2,457	431	.784	361	.657

FLAX MACHINERY.

Draught.	Gills Speed.	Rolls Speed.	Spindle Speed.	Ft. Lb. per Second.	Horse-Power.
8. to 1	180 per min.			340	0.619
7. " 1	180 "			632	1.149
8.½ " 1		100 Revol'ns,	590	1,297	2.358
6. " 1		30 "	2,665	1,860	3.382
30. " 1	54 per min.	70 "		560	1.018
16. " 1	108 "			427	.767
14. " 1	102 "	84 "		604	1.097
12. " 1	130 "	108 "	540	1,132	2.058
7.89 " 1		39 "	2,925	2,335	4.246
7.89 " 1		40 "	3,176	2,702	4.913
				520	.947
				790	1.436
				1,077	1.957
			2,700	1,903	3.460
				1,653	3.005
				2,853	5.186

WOOL MACHINERY.—(Continued.)

WOOLEN-LOOMS.

Date.	Place.	Description.	Width.	Harness	'Picks' per min.	Ft. Lbs.	H. P.
June, 1871	Derry Mills,	1 Satinet Loom.	36 in.		95	122	.221
"	"	{ 1 Broad Crompton Loom,	90 "	10	86	285	.519
Oct., 1871	{ Manchest. Print Works.	1 " "	90 "	10	86	348	.633
"	"	1 Thomas,	108 "	10	65	233	.424

FINISHING-MACHINERY.—(Continued.)

Date.	Place.	Description.	Revolu.	Ft. Lb.	H. P.
June, 1871	Derry Mills,	1 Broad Gig, 54 inches, empty,	160	279	.507
"	"	" set light, old Teazles,	160	570	.927
"	"	" set heavy, new "	160	1,100	2.000
"	"	1 Rotary Fulling-Mill, [ton & Co.,	126	1,395	2.536
"	"	1 38 in. Hydro-Ex., Rice, Barton & Co.,	580	1,100	2.000
"	"	1 30 in. Hydro-Ex., Laconia Pat.,	630	999	1.817

TESTS OF SHAFTING.

Date.	Place.	Length.	Diameter.	Weight.	Weight of Pulleys.	Total Weight.
April 1871	Amoskeag Mills,	8 ft. 6 in.	2 1/8 inc.	101 lb.	577 lb.	678 lb.
"	"	34	"	404	1,974	2,378
"	"	114	"	1,366	1,859	3,225
"	"	228	"	2,732	3,617	6,349
"	"	342	"	4,098	5,331	9,429
"	"	16	2 5/8 inc.	2,427	2,988	5,415
"	"	178	2 1/8 "			
"	"	10 ft. 4 in.	4 1/8 "			
"	"	80	2 7/8 "			
"	"	32	2 5/8 "	3,910	5,393	9,303
"	"	48	2 3/8 "			
"	"	32	2 1/8 "			
July, 1871	"	10	2 5/8 "			
"	"	48	2 3/8 "	1,289	1,456	2,745
"	"	32	2 1/8 "			
"	"	Similar line			1,006	2,295
"	"	34	2 5/8 "			
"	"	32	2 3/8 "	1,484	1,736	3,220
"	"	32	2 1/8 "			
"	"	10 ft. 4 in.	2 7/8 "	2,336	2,999	5,335
"	"	176	2 1/8 "			
"	"	Similar line		2,336	2,999	5,335
Jan., 1872	Whittenton Mills,	Ab't 200 ft.	2 1/4 "	2,700*	.....	......
Feb., 1872	Langdon "	24 ft. 10 in	2 1/8 "	295	350	645
Mar., 1872	Haydensville,	42	2 1/8 "	428	.....	......
April 1, '72	Salmon Falls, N. H.,	9	4 "	3,151	2,354	5,805
		231	2 1/8 "			
April 2, '72	"	Same shaft.		3,151	2,354	5,805
May, 1872	Rockport, Mass.	9 ft. 10 in	3 "			
	"	32	2 3/8 "	1,987	2,000*	3,987
	"	32	2 1/8 "			
		96	1 7/8 "			
May, 1872	Masconom't M'l Newburyport,	10 ft. 8 in.	3 "			
		24	2 1/2 "			
		127 ft. 1 in.	2 1/4 "	3,554	4,268	7,882
		64	2 "			
		64 ft. 3 in.	1 3/4 "			
Nov., 1872	Paterson, N. J.,	100	1 5/8 "	unkn'wn	unkn'wn	unkn'wn
Nov., 1873,	Granite Mills, Fall River,	Ab't 200 ft.	2 "	"	"	"

* Estimated.

TESTS OF SHAFTING.

No. of Bear'gs.	Rev. per Minute.	Ft. Lbs.	Horse-Power.	Coeff. Friction	REMARKS.
					CONTINUOUS OILING. [Kerosene Oils mixed·
2	216	49	.089	.0336	Single Counter. Dreyfuss Pat. Oiler. Sperm and
8	216	196	.357	.0413	4 Count's like above. Connected with Belts. Same
15	216	325	.590	.0500	Single line. Oilers as above. [Oils mixed·
30	216	650	1.181	.0501	2 Lines like above connected. Oils, etc., same.
47	216	1,022	1.858	.0552	3 " " " " "
25	216	378	.687	.0338	Single line, " "
26	210	873	1.587	.0334	Single line, Oils, etc., same.
					ORDINARY OILING.
12	150	273	.499	.0640	Single line. Oiled in ordinary way, daily. Tallow in Boxes, as safeguard in case of Heating.
12	150	217	.394	.0610	" " " "
13	150	291	.537	.059	" " " "
24	211	793	1.442	.0759	Single line. Had been oiled A. M., test at 11 A. M.
24	211	679	1.234	.0650	" " Taken just after oiling.
....	155	314	.571	......	
4	210	143	.260	.114	" " Sprung in centre by pull of Belt.
5	120	147	.267	......	
31	211	857	1.558	.0714	" " Tallow in Boxes, taken at noon; had been oiled early in A. M.
31	211	619	1.120	.0516	Tallow removed from Boxes, and sponge saturated in oil substituted. Time of testing as before.
30	185	376	.685	.0568	Oiled daily in usual manner.
37	245	1,028	1.870	.0585	" " " "
11	170	184	.335	......	Dreyfuss Oilers.
....	200	302	.549	......	" "

TESTS FROM 1874-'79.

POWER OF COTTON

Date.	Place.	Machine.	No. of Beaters.	Diam. Beaters.	Rev. per Minute.
Jan., 1874	Clipper Mills, Baltimore, Md.	Kitson Lapper, Whitehead & Atherton, Old Pattern,	3 3	16 in. 18 in.	1,380 1,380
Feb., 1874	Jackson Co., Nashua,	Kitson's Compound,	4	2. 12 in. 2. 16 in.	1,380
"	Boott Mills, Lowell,	" " New Pattern, "Broken Beaters,"	4	1. 9 in. 1. 12 in. 2. 16 in.	700 950 1,380
"	Whittenton Mills, Taunt'n,	Kitson's Compound, Regular Pattern,	4	2. 16 in. 2. 12 in.	1,390
Mar., 1874	"	Same Machine, 1st pair Beaters removed, and Whitehead Whipper put in place,	3	1. 24 in. 2. 16 in.	1,000 1,390
"	"	Same Machine, Toothed Beater substituted by Kitson,	3	1. 24 in. 2. 16 in.	1,390
"	"	Same Machine, 4 Bladed Beater in place of Tooth'd Btr.,	3	1. 24 in. 2. 16 in.	933 1,390
April, 1874	Stark Mills, Manchester,	1. Kitson Compound, Regular Pattern,	4	2. 12 in. 2. 16 in.	1,390
"	"	"	"	"	1,390
May, 1874	"	Same Machine,	"	"	1,400
"	"	Similar,	"	"	1,150
Aug., 1874	Nashua Mfg. Co.,	Whitehead & Atherton Compound, 1 Whipper, 1 Beater,	2	1. 24 in. 1. 16 in.	1,130 1,500
Mar., 1875	China Mills, Suncook,	1. Kitson Compound,	4	2. 12 in. 2. 16 in.	1,450
"	Webster Mills, Suncook,	1. " "	3	16 in.	1,300
"	China Mills, Suncook,	1. Whitin Lapper,	3	12 in.	1,700
July, 1875	Appleton Mills,	Whitehead & Atherton Opener,	2	1. 24 in. 1. 16 in.	1,080 1,380
Nov., 1875	Boott Mills,	Kitson's Atmospheric Opener,	2 Disks 2 Beaters	16 in. 16 in.	1. 620 1. 666 1.1,200 1.1,430
Aug., 1876	Prescott Mill,	Kitson's New Feed Motion,		....	240

PICKERS AND LAPPERS.

No.of Fans.	Rev. per Minute.	Wt. Lap per Yd.	Lbs. per Day.	Ft. lb. Power.	Horse-Power.	Remarks.					
3	1,380	16 oz.	3,930	4,380.53	7.969	Opening from Bale. Without cotton, 5,812 H.P.					
3	1,380	12 oz.	3,600	5,883.33	10.607	"	"	"	"	"	7.969 H.P.
3	1,380	13 oz.	3,840	5,674.70	10.318	"	"	"		"	8.346 H.P.
3	1. 1,730 2. 1,380	14 oz.	4,800	4,525.64	8.228	"	"	"	"	"	5,950 H.P.
3	1,390	15 oz. 15 oz.	3,300 3,350	8,142.86 6,308.31	14.805 11.505	Carded & Dyed Cotton. " " " "				"	9.177 H.P.
3	1. 1,000 2. 1,390	19½ oz.	4,420	5,000	9.091	Opening from Bale.				"	7.102 H.P.
3	1. 1,000 2. 1,390	15½ oz.	3,350	6,588	11.978	Carded & Dyed Cotton. "				"	6.711 H.P.
3	933 1,390	15 oz.	3,320	7,521	13.678	"	"		"	"	8.052 H.P.
3	1,390	17 oz.	4,000	6,766	12.302	Cotton from Bale,				"	10.732 H.P.
3	1,390	13 oz.	3,060	6,394	11.623	"	"	"			
3	1,400	17 oz.	4,000	7,208	13.105	"	"	"	"	"	11.139 H.P.
3	1,150	11 oz.	2,220	4,800	8.725	"	"	"	"	"	7,027 H.P.
2	1. 1,700 1. 1,500	....	5,370 5,000	4,349 4,814	7.907 8.754	Previously opened Cotton. Cotton from Bale.	"			"	5.793 H.P.
3	1,450	18 oz.	4,000	6,300	11.453	"	"	"			
2	1,300	8 oz.	2,700	3,361.61	6.112	"	"	"		"	5.442 H.P.
3 2	1,400 1,380	7 oz. 12 oz 14 oz.	1,600 3,000 3,400	2,844.44 3,219.51 3,463.42	5.172 5.854 6.3	As running. Cotton from Bale. Fed heavier on Aprons.				"	4.05 H.P.
3	1,500 1,480 1,440	16 oz.	4,000	4,200	7.636	Cotton from Bale.	"				6.856 H P.
....	....	....	4,000	193.85	.352	Tested separate from Picker.					

COTTON CARDS.

Date.	Place.	Maker.	Width.	Rev. per Minute.
March, 1875	China Mills, Suncook,	Wm. Mason Breaker,	36 in.	127
" "	" " "	" " Finisher,	36 in.	127
" "	Pembroke Mills, "	" " Breaker,	36 in.	127
" "	" " "	" " Finisher,	36 in.	127
" "	Webster Mills, "	" " Breaker,	36 in.	127
" "	" " "	" " Finisher,	36 in.	127
April, "	Newton, Mass.,	O. Petter "	36 in.	128
August, 1876	Mass. Mills, Lowell,	Lowell Ma. Shop Finisher,	36 in.	125
" "	" " "	" " " Breaker,	36 in.	125
" "	" " "	Foss & Pevey, Single,	36 in.	125
April, 1877	Lockwood Mills,	Saco W. P. Co. Breaker,	36 in.	133
" "	Waterville, Maine,	" " " Finisher,	36 in.	133
" "	" "	" " " Single,	36 in.	133

RAILWAYS FOR CARDS.

Date.	Place.	Maker.
March, 1875	China Mills, Suncook,	Wm. Mason Breaker,
" "	" " "	" " Finisher,
" "	Webster " "	" " "
" "	Pembroke " "	" " "
" "	Newton, Mass.,	" " "
August, 1876	Mass. Mills, Lowell,	Lowell Finisher,
" "	" " "	" Foss & Pevey Finisher,
April, 1877	Lockwood Mills, Waterville,	" Breaker,
" "	" " "	" Finisher,

COTTON DRAWING-FRAMES.

Date.	Place.	Maker.	No. Rolls.
March, 1875	China Mills, Suncook,	Mason, 1st,	4
" "	" " "	" 2d,	4
" "	Webster " "	" 1st,	4
" "	" " "	" 2d,	4
" "	Pembroke " "	" 1st,	4
" "	" " "	" 2d,	4
" "	Newton Mills, Newton, Mass.,	Whitin, 1st,	4
" "	" " " "	" 2d,	4

COTTON CARDS.

Lb. per Day.	Ft. Lbs. each.	H. P. each.	Cards per Railway.	H. P. Railway.	No. of Cards per H P., including Railway.
45	149.60	.272	40	1.361	3.27
46	149.60	.272	9	.659	2.90
45	145.65	.265	40	1.360	3.34
46	145.65	.265	9	.424	3.20
45	180	.327	..	....	....
45	137.5	.250	9	.521	3.25
53	80	.145	14	.846	4.86
65	146.08	.266	8	.473	3.07
65	159.77	.283	..	....	....
65	192.77	.350	8	.554	2.38
48	97	.177	72	2.085	4.85
48	75.60	.137	12	.546	5.48
27	61.92	.112	12	.546	6.35

RAILWAYS FOR CARDS.

No of Cards.	Diam. F. Roll.	Rev. F. Roll.	Draft.	Wt. Sliver.	Ft. Lbs. per Sec.	H.-Power.
40	12 in.	12	....	8 oz.	748.91	1.361
9	1¼ in.	445	4.21	95 gr.	362.5	.659
9	1¼ in.	370	3.50	89 gr.	286.86	.521
9	1¼ in.	340	3.50	89 gr.	233.33	.424
14	1¼ in.	570	3.50	120 gr.	465.22	.846
8	1¼ in.	250	....	......	505	.554
8	1¼ in.	350	....	......	260	.473
72	12 in.	12	....	......	1,145.83	2.085
12	1¼ in.	300	4	90 gr.	300.3	.546

COTTON DRAWING-FRAMES.

Rev. Rolls.	Doublings.	Draft.	No. of Deliveries.	Ft. Lb. per Delivery.	H. P per Delivery.	Total H. P.
340	4	4.24	4	126.13	.229	.917
348	2	4.08	8	63.06	.114	.917
280	4	4.12	4	106.25	.194	.776
570	2	4.12	8	75	.135	1.091
310	4	4.50	3	141.02	.25	.769
380	2	4.50	6	86.24	.157	.941
312	3	4	8	80.60	.147	1.173
390	3	4	6	123.50	.224	1.345

ROVING FRAMES.

Date.	Place.	Makers.	Size of Bobbin.	No. Spin.	Rev. Spin.	Diam. of Roll.
Aug., 1874	Amoskeag Co.,	Howard & Bullough,	10 × 5	56	620	$1\frac{1}{8}$
" "	Manchester,	" "	9 × 4.5	68	602	"
" "	"	" "	7 × 3.5	116	930	"
Jan., 1877	Conant Thread Co.,	Higgins & Sons,	5 × 2.5	140	1120	"
" "	Pawtucket,	Prov. Machine Co.,	"	148	1260	"
April, "	Lockwood Mills,	Saco W. P. Co.,	11 × 5.5	64	700	$1\frac{1}{4}$
" "	Waterville, Maine,	"	10 × 4	88	855	$1\frac{1}{8}$
" "	" "	"	7 × 3.5	160	1175	"
Dec., "	Whittenton Mills,	Prov. Machine Co.,	"	128	1100	"
" "	" "	Wm. Mason,	"	"	900	"

DEAD SPINDLE

Date.	Place.	Maker.	Size of Bobbin.	No. Spin.	Rev. Spin.	Diam. of F. Roll.
Sept., 1873	Atlantic Mills,	Lowell Shop Speeder	12 × 6	34	625	$1\frac{1}{4}$
" "	Lawrence,	" " "	12 × 6	34	600	$1\frac{1}{4}$
" "	"	" " "	10.5 × 4.5	36	750	$1\frac{1}{4}$
" "	"	" Intermediate	9.5 × 4.5	58	900	$1\frac{3}{16}$
" "	"	" "	9.5 × 4.5	58	900	$1\frac{3}{16}$
" "	"	" "	9.5 × 4.5	58	900	$1\frac{3}{16}$
" "	"	" Fine Speeder	8.5 × 3.5	72	1200	$1\frac{1}{8}$
" "	"	" " "	8.5 × 4.5	50	900	$1\frac{1}{8}$
" "	"	" Intermediate	9.5 × 4.5	58	900	$1\frac{3}{16}$
Mar., 1875	China Mills,	O. Pettee, 1st,	9 × 4.5	30	800	$1\frac{1}{4}$
" "	Suncook,	" 2d,	7 × 3	78	1420	$1\frac{1}{8}$
" "	Webster Mills,	" 1st,	8 × 4	28	820	$1\frac{1}{4}$
" "	Suncook,	" 2d,	$6\frac{1}{2}$ × 3	78	1500	$1\frac{1}{8}$
" "	Pembroke Mills,	" 1st,	8 × 4	24	800	$1\frac{1}{4}$
" "	Suncook,	" 2d,	6 × 3	72	1420	$1\frac{1}{8}$
" "	"	" 1st,	9 × 4	40	800	$1\frac{1}{4}$
" "	Newton Upper Falls,	" 1st,	9 × 4	36	864	$1\frac{1}{4}$
" "	Newton, Mass,	" 1st,	10 × 5	36	593	$1\frac{1}{4}$

ROVING FRAMES.

Rev. Roll.	Draft.	From	To	Ft. Lb. Frame.	Ft. Lb. Spindle.	H. P. Frame.	Spindles per H.P.	Notes.	
193	4.57	77. g.	0.52bk	639.99	11.43	1.163	48.1	Average ¼ full.	
143	4	0.52 hk	1.04	422.93	6.22	.768	88.4	"	"
120	6	1.04 "	3	636	5.48	1.156	100	"	"
80	8	5 "	20	554.54	3.96	1.008	140	"	"
82	8	5 "	20	393.33	2.66	.715	207	"	"
168	4	.55 gr.	0.60	820	12.81	1.491	43	"	"
145	5	0.60 hk	1.50	820	9.31	1.491	59	"	"
129	5.80	1.50 "	4.70	930.43	5.81	1.701	94.5	"	"
175	"	2 "	....	1290	10.21	2.345	55	"	" New Frame.
147	"	2 "	....	760	5.94	1.382	93	"	" Old Frame.

ROVING FRAMES.

Rev. F. Roll.	Draft.	From	To	Ft. Lb. Frame.	Ft. Lb. Spindle.	H. P. Frame.	Spindles per H.P.	Notes.		
198	3.06	89 g.	0.30bk	541.66	16	.984	34	Bobbin ½ full.		
185	4.54	"	0.49	441.41	13	.800	42.5	"	"	
152	4.30	"	0.45	472.44	13.12	.859	42	"	"	
190	6.04	0.49 hk	1.25	704	12.14	1.280	45.30	"	" as running.	
190	6.04	"	1.25	670	11.67	1.219	47.25	"	" freshly oiled.	
190	6.04	"	1.25	700	12.08	1.273	45.50	"	" similar frame as	
173	3.68	1 "	1.81	863	12	1.570	46	"	" as run'g. [r'g.	
116	7.75	0.45 "	1.81	512.29	10.24	.931	53.60	"	" as running.	
200	5.78	0 30 "	1	576	10	1.048	55	"	"	"
166	5.97	50 g.	1.25	388.23	12.94	.706	42.5	"	"	"
127	7.03	1.25 "	3.71	1200	15.38	2.182	35.76			
170	5.97	45 g.	1.25	465.12	16.61	.846	32	"	"	"
148	7 03	1.25	3.71	1480.52	18.97	2.692	29	"	"	"'
166	5.97	45 gr.	1.25	512.5	21.35	.932	26	"	"	"
140	8.03	1.25 "	4.12	1333.33	18.52	2.484	30	"	"	"
150	5.32	45 gr.	1.20	712.50	17.87	1.297	31	"	"	"
157	12.82	88 "	1.50	529.41	14.73	.963	37.4	"	"	"
153	8.32	100 gr.	1.50	342.31	9.51	.622	57.8	"	" New Pattern.	

RING SPINNING.

Date.	Place.	Description.	Diam. Ring.	No. Spin.	Weight Spindle.	Rev. Spin.	Rev. Roll.
Mar., 1874	Whittenton Mills, Taunton, Mass.,	Wm. Mason,	1$\frac{3}{4}$ in.	160	12 oz.	5,000	110
" "	"	" "	"	160	"	5,000	110
June, "	Lawrence Manuf. Co.,	Lowell M. Shop,	1$\frac{5}{8}$	224	12$\frac{1}{4}$	5,095	82
Jan., 1875	China Mills, Suncook, N. H.,	Wm. Mason,	"	128	14	5,000	61
" "	" "	" "	"	128	"	4,500	56
Feb., "	Renfrew Co., S. Adams,	Whitin,	"	144	12	5,000	96
Mar., "	China Mill, Suncook,	Wm. Mason,	"	128	14	5,160	66
" "	" " "	" "	"	128	"	5,000	64
" "	Webster Mill, Suncook,	" "	1$\frac{1}{2}$	128	"	5,000	64
" "	" " "	" "	"	128	"	5,160	66
" "	" " "	" "	"	128	"	5,160	66
" "	" " "	" "	"	128	"	5,160	66
" "	" " "	" "	"	112	"	5,000	64
" "	" " "	" "	"	112	"	5,075	65
" "	Pembroke Mill, "	Saco W. P. Shop	"	112	12	5,080	58
" "	" " "	" " "	"	112	"	5,080	58
" "	" " "	" " "	"	112	"	5,080	58
" "	Newton Falls, Mass.,	" " "	1$\frac{5}{8}$	128	"	5,120	68
" "	" " "	" " "	"	128	"	5,120	67
" "	" " "	" " "	"	128	"	5,060	66
" "	" " "	" " "	"	128	"	5,060	66
" "	" " "	" " "	"	128	"	5,120	64
" "	" " "	" " "	"	128	"	5,120	72
April, "	Manchester Mills,	" " "	"	192	11$\frac{1}{2}$	5,757	71
" "	" "	" " "	"	192	"	5,757	..
" "	" "	" " "	"	128	"	5,700	61.5
" "	" "	" " "	"	128	"	5,700	61.5
" "	" "	" " "	"	128	"	5,700	61.5
" "	" "	" " "	"	128	"	5,700	61.5
July, "	Pacific Mills,	Whitin,	"	160	12	5,030	63
" "	" "	"	"	160	"	5,030	61.5
" "	" "	"	"	160	"	5,030	61.5
" "	" "	"	"	160	"	5,030	61.5
April, 1876	York Mills, Saco,	Saco W. P. Shop	"	128	11$\frac{3}{4}$	5,800	82
" "	" " "	Saco W. P. Sh. Old Throstle,	....	128	flier, 3.40	3,610	60
June, "	Everett Mills, Lawrence	Lowell Sh., O. T.	....	128	"	3,750	57
April, 1877	Clinton, Woonsocket,	Whitin,	"	128	12	5,670	71
" "	" "	"	"	128	"	5,670	71
Dec., 1875	Jackson Co., Nashua,	Lowell Throstle,		192	flier, 4 oz.	3,260	93
Nov., 1877	Atlantic Mills,	" Shop Ring	"	176	13$\frac{1}{8}$	5,022	88
Dec., "	Whittenton Mills,	Wm. Mason,	1$\frac{3}{16}$	128	12	4,800	128
Jan., 1878	Amoskeag Mills,	Amoskeag Co.,	1$\frac{1}{2}$	128	11	3,910	89
June, "	Lyman Mills No. 1 Holyoke,	Whitin,	1$\frac{5}{8}$	128	12	4,200	132
" "	"	"	"	128	"	4,200	132
" "	"	Whitin, alter'd to light Spindle No. 1.	1$\frac{7}{16}$	128	7$\frac{3}{4}$	5,593	116
" "	"	"	"	128	"	5,593	116
" "	"	"	"	128	"	5,593	116
" "	"	No. 2, different form of spindle	"	128	"	5,593	116
" "	"	"	"	128	"	5,593	116
" "	"	"	"	128	"	5,593	116

ORDINARY SPINDLE.

Draft.	Rov'g.	Yarn.	Ft. Lb. Frame.	Ft. Lb. Spin.	H. P. Fr'me.	Spindles per H.P.	Remarks.
7	2	14 warp.	1,212.50	7.58	2.205	72.5	Average half full. Heavy Bobbin, 1 oz.
7	2	14 "	1,149.37	7.18	2.090	76.6	" " " Light Bobbin, ¼ oz.
8	2.25	18 "	1,324.56	5.93	2.405	93	" " " Hussey's Banding.
7.92	3.54	28 "	878.12	6.86	1.596	80.2	" " "
7.92	3.54	28 "	638.24	4.99	1.160	110	" " "
8	2	15 "	977.28	6.75	1.717	81.5	" " "
7.72	3.71	28.50"	857.14	6.70	1.543	83	" " " Bands rather tight.
7.72	3.71	28.50"	763.16	5.96	1.387	92.5	" . " " " easy.
7.72	3.71	28.50"	941.25	7.35	1.711	75	" " " " tight.
7.72	3.71	28.50"	1,137.37	8.69	2.196	62.4	" " " " very tight.
7.72	3.71	28.50"	1,540.62	12.03	2.801	46	" " " New B'ds, " " indeed.
7.72	3.71	28.50"	915.62	7.15	1.665	77	" " " " " easy.
7.72	3.71	28.50"	875	7.81	1.591	70.5	" " " Bands aver. tight. of Mill.
7.72	3.71	28.50"	894.25	7.98	1.625	69	" " " " " "
7.72	3.71	28.50"	610.46	5.45	1.110	101	" " " " " "
7.72	3.71	28.50"	69'.86	6.17	1.258	90	" " " Similar Fr'me, B'ds tight.
7.72	3 71	28.50"	622.60	5.55	1.131	99	" " " " " cased.
6.78	3.62	28 "	669	5.23	1.216	105	" " " Bands properly adjusted.
6.78	3.62	28 "	677.6	5.30	1.232	104	" " " Sim. F'me, B'ds prop. ad.
6.78	3.62	28 "	785.72	6.12	1.429	90	" " " " " Bands tight.
6.78	3.62	28 "	675.33	5.28	1.228	104	" " " Same " " adjusted.
6.78	3.62	28 "	674.25	5.26	1.226	104	Bobbin ¼ full. Bands in good order.
6.78	3.62	28 "	703.9	5.50	1.280	100	" " Rolls higher speed.
7.66	4	30 "	1,049.24	5.46	1.908	101	" "
....	....	......	757.7	3.94	....	....	" empty. Rolls stopped. Spindle only.
7.68	11 dbl	40 "	735.07	5.75	1.336	95.5	" ¼ full. Taken as running.
7.68	11	40 "	695.92	5.60	1.303	98.4	" " Bands adjusted.
7.68	11	40 "	750	5.86	1.364	94.8	" " Spindle shortened ¼ inch.
7.68	11	40 "	734.33	5.74	1.335	96	" " Spin. short., B'ds adj., 2d trial.
6.94	4.40	28.50"	793.75	4.96	1.443	111	" " Taken as running.
6.94	4.40	28.50"	843.75	5.27	1.534	104	Full Bobbin. " "
6.94	4.40	28.50"	862.50	5.39	1.568	102	" " Bobbin ⅜ in. longer.
6.94	4.40	28.50"	931.25	5.82	1.693	94.5	" " " Damp mor.
7	3	21 "	976.19	7.62	1.775	72	Av'ge ¾ full. Bands a litt'e tight.
7	3	21 "	898.33	7.01	1.634	78	" " { Bands in good order. Tested for comparison.
7.50	3	22.50"	1,059.07	8.27	1.928	66.30	" " "
8.14	3.68	32 "	700	5.47	1.272	100.5	" " Taken as running.
8.14	3.68	32 "	637.50	4.98	1.159	110.4	" " Draper's "Tension Regulator."
9	1.56	13.50w'ft	1,076 .	5.60	1.957	98.21	" " Tested for comparison.
7.60	2	15 warp.	1,160.38	6.60	2.110	83.50	" "
10	5 dbl	9 "	1,017.26	7.94	1.849	70	" " Bands rather tight.
7	2	14 "	683.91	5.34	1.223	103	" " " easy.
....	....	13 weft.	537.38	4 20	.977	Ave'ge	Empty Bobbin } Speed variable.
....	....	13 "	528.03	4.11	.960	132+	Full " }
....	....	23 "	576.92	4.51	1.049	....	Empty " } Full Bobbin tested first, then empty, then ¼ full, the last probably correct.
....	....	23 "	524.04	4.18	.958	131.5	Half full " }
....	....	23 "	581.73	4.23	1.053	......	Full " }
....	....	23 "	692.30	5.41	1.259	......	Empty " } Same notes as above.
....	....	23 "	629.80	4.92	1.145	112—	Half full " }
....	....	23 "	701.92	5.48	1.276	......	Full " }

RING SPINNING.—(Continued.)

Date.	Place.	Maker.	Diam. Ring.	No. Spin.	Weight Spindle.	Rev. Sp.	Rev. F. Roll.
June, 1878	Lyman Mills No. 1, Holyoke,	No.1 Spindle, Coarser Y'n,	$1\frac{7}{16}$ in.	128	$7\frac{1}{4}$ oz.	4,033	135
" "	"	"	"	128	"	4,033	135
" "	"	"	"	128	"	4,033	135
" "	"	No. 1, Higher Speed,	"	128	"	6,282	132
" "	"	"	"	128	"	6,282	132
" "	"	"	"	128	"	6,282	132
" "	"	Whitin Lt. Sp.,	$1\frac{7}{8}$	128	10	5,383	121
" "	"	"	"	128	"	5,383	121
" "	"	"	"	128	"	5,383	121
" "	No. 2 Mill,	"	$1\frac{3}{8}$	176	6	7,400	62
Oct., 1878	Prescott Mill,	Lowell Shops, Reg. Sp.,	$1\frac{5}{8}$	176	12	5,600	86
" "	"	"	"	176	"	5,600	86
" "	"	"	"	176	"	5,600	86
" "	"	"	"	176	"	5,600	86
" 15, "	Pacific Mills, Lawrence,	Whitin Spindle, Cut-off at Butt,	"	160	7	6,059	73
" " "	"	"	"	160	"	6,059	73
" " "	"	"	"	160	"	6,059	73
" " "	"	"	"	160	"	5,073	61
" " "	"	"	"	160	"	5,073	61
" " "	"	"	"	160	"	5,073	61
" 22, "	"	"	"	160	"	6,059	73
" " "	"	"	"	160	"	6,059	73
" " "	"	"	"	160	"	6,059	73
" 25, "	"	"	"	160	"	6,059	73
" " "	"	"	"	160	"	6,059	73
" " "	"	"	"	160	"	6,059	73
" 16, "	"	Regular Whitin Spindle, Old Bolster,	"	160	12	5,063	61
" " "	"	"	"	160	"	5,063	61
" " "	"	"	"	160	"	5,063	61
" " "	"	Similar Fr., Coned-up Bolster,	"	160	"	5,063	61
" " "	"	"	"	160	"	5,063	61
" " "	"	"	"	160	"	5,063	61
" 22, "	"	Same Frame,	"	160	"	6,059	73
" " "	"	"	"	160	"	6,059	73
" " "	"	"	"	160	"	6,059	73
" 23, "	"	"	"	160	"	6,059	73
" " "	"	"	"	160	"	6,059	73
" " "	"	"	"	160	"	6,059	73

ORDINARY SPINDLE.—(Continued.)

Draft.	Rov'g.	Yarn.	Ft. Lb. Frame.	Ft. Lb. Spin.	H. P. Frame	Spindle per H.P.	Remarks.		
.....	.....	13 weft	435.19	3.4	.791	Av'ge 163	Empty Bobbin. Half full " Full "	Speed irregular as before.	
.....	.....	13 "	440.37	3.44	.8				
.....	.....	13 "	420.56	3.28	.765				
.....	.....	23 "	656.86	.....	1.194	110+	Empty " Half full " Full "	Speed more even.	
.....	.....	23 "	666.67	4.99	1.212				
.....	.....	23 "	676.47	.....	1.24				
.....	.....	12 warp	733.33	5.73	1.333	Av'ge 85.4	Empty " Half full " Full "	Speed irregular.	
.....	.....	12 "	850	6.64	1.545				
.....	.....	12 "	888.88	6.94	1.616				
.....	.....	60 "	1,118.42	6.35	2.033	86.6	Half full "		
.....	.....	21 "	950.82	5.4	1.727	Av'ge 96	Empty " Full "	Pearl Bobbin, 6 in. Traverse. 842.18 Grains Yarn.	
.....	.....	21 "	1,065.57	6.05	1.938				
.....	.....	21 "	950.82	5.4	1.727	Av'ge 96.8	Empty " Full "	Common Bobbin, 5¼ in. Trav. 645.75 Gr. Yarn.	
.....	.....	21 "	1,049.80	5.96	1.909				
6.79	4.40	28.50 "	1,021.28	6.38	1.857	86	Empty "	Wt. Bob., 847.25 gr. Yarn, 527.75 gr. Taken as found.	
6.79	4.40	28.50 "	1,191.49	7.447	2.166	73.9	Full "		
6.79	4.40	28.50 "	1,106.38	6.91	2.011	80	Average "		
6.79	4.40	28.50 "	718.18	4.49	1.37	122	Empty "	Lower speed, as ordinarily run.	
6.79	4.40	28.50 "	845.46	5.28	1.537	104	Full "		
6.79	4.40	28.50 "	781.81	4.89	1.453	113	Average "		
6.79	4.40	28.50 "	1,031.91	6.45	1.876	85.3	Empty Bob.	Rings centred and Bobbins selected. Wt. Bob., 871.9 gr. Wt. Yarn, 541.4 gr.	
6.79	4.40	28.50 "	1,202.13	7.51	2.186	73.2	Full "		
6.79	4.40	28.50 "	1,117.02	6.98	2.031	78.8	Average "		
6.79	4.40	28.50 "	978.72	6.12	1.78	89.9	Empty "	Rings and Bobs. as above. Wt. reduc'd 10 p.ct. Weather very dry and clear. Wt. still further reduc'd 20 p.ct. without ch. in power req'd.	
6.79	4.40	28.50 "	1,148.94	7.18	2.089	76.6	Full "		
6.79	4.40	28.50 "	1,063.83	6.65	1.934	82.7	Average "		
6.79	4.40	28.50 "	807.23	5.045	1.468	109	Empty Bobbin.	Taken as running. Usual speed.	
6.79	4.40	28.50 "	879.52	5.5	1.6	100	Full "		
6.79	4.40	28.50 "	843.37	5.27	1.534	104.5	Average "		
6.79	4.40	28.50 "	819.28	5.12	1.49	107.4	Empty "	Taken as running. Usual speed.	
6.79	4.40	28.50 "	891.57	5.57	1.621	98.8	Full "		
6.79	4.40	28.50 "	855.42	5.345	1.555	103.1	Average "		
6.79	4.40	28.50 "	1,014.29	6.34	1.844	87	Empty "	Speed increased. Wt. Bobbin, 871.9 gr. " Yarn, 456.7 gr.	
6.79	4.40	28.50 "	1,157.14	7.23	2.014	76	Full "		
6.79	4.40	28.50 "	1,085.71	6.786	1.909	81.5	Average "		
6.79	4.40	28.50 "	1,071.43	6.7	1.948	82	Empty "	Damp morning. Storm of 24th coming on. Bob. & Y'n as before.	
6.79	4.40	28.50 "	1,200	7.5	2.182	73.3	Full "		
6.79	4.40	28.50 "	1,135.71	7.1	2.065	77.7	Average "		

RING SPINNING.—(*Continued.*)

Date		Place.	Maker.	Diam. Ring.	No. Spin	Weight Spindle.	Rev. Spindle.	Rev F. Roll.
June,	1874	Lawrence Co., Lowell, Mass.,	Lowell M. Shop,	$1\frac{5}{8}$ in.	224	4 oz	6,400	103
"	"	"	"	"	224	"	6,340	102
"	"	"	"	"	224	"	6,340	102
Sep.,	"	"	"	"	224	$5\frac{1}{4}$	6,540	98
Nov.,	"	Stark Mills, Manchester, N. H.,	Altered Throstle,	$1\frac{7}{8}$	128	"	5,130	110
"	"	"	"	"	128	"	5,130	110
Jan.,	1875	China Mill, Suncook, N. H.,	Altered Mason,	$1\frac{5}{8}$	128	"	5,800	70
"	"	"	Ring Frame,	"	128	"	6,000	72
Feb.,	"	Renfrew Mfg. Co., South Adams,	Whitin, altered,	"	144	"	5,380	98
"	"	"	"	"	144	"	6,050	110
Aug.,	"	Stark Mills, Manchester,	Lowell M. Shop,	2	128	$5\frac{7}{8}$	6,070	117
"	"	"	New Frames,	$2\frac{1}{4}$	128	"	5,976	116
Oct.,	"	Hamilton Co., Lowell,	"	$1\frac{5}{8}$	208	4	5,600	90
"	"	"	"	"	208	"	5,600	90
Jan.,	1876	Salmon Falls, N. H.,	Altered Frame,	"	144	"	5,800	74
"	"	"	Throstle,	"	144	"	6,050	77
April,	"	York Mills, Saco,	"	"	128	$5\frac{1}{4}$	5,800	82
"	"	"	"	"	128	"	5,800	st'p'd
June,	"	Everett Mills, Lawr'ce,	"	"	128	5	3,800	58
"	"	"	"	"	128	"	4,885	72
Aug.,	"	Boott Mills,	Lowell Shop,	"	224	$5\frac{1}{2}$	6,020	97
April,	1877	Clinton Mills, Woonsocket,	Whitin, alt'd to Mod. Pat., Sawyer Spin.	"	192	$4\frac{1}{4}$	7,500	96
"	"	"	"	"	192	....	7,000	st'p'd
"	"	Lockwood Mills, Waterville, Me.,	Saco W. P. Co. Shop,	"	176	$5\frac{1}{4}$	7,000	90
"	"	"	"	"	176	"	7,000	90
"	"	"	"	"	176	"	7,300	105
Nov.,	"	Jackson Co., Nashua,	Lowell M. Shop,	$1\frac{3}{4}$	192	$4\frac{3}{4}$	5,568	116
"	"	"	"	"	192	"	5,568	116
Dec.,	"	Whittenton Mills,	Wm. Mason, Old, alt'd to Sawyer Spin.	$1\frac{13}{16}$	128	"	5,135	130
"	"	"	"	"	128	"	5,135	130
Jan.,	1878	Amoskeag No. 4,	Whitin M. Co.,	$1\frac{3}{8}$	144	$4\frac{1}{4}$	6,230	118
"	"	"	Drap'r's Fil'g Fr., with Mod. Sawyer Spin.	"	144	"	6,230	118
"	"	"	"	"	144	"	5,925	112
"	"	"	"	"	144	"	5,234	104
"	"	"	"	"	144	"	4,960	120
"	"	"	"	"	144	"	3,658	108
Mar.,	"	Nashua Mfg. Co.,	Lowell M. Shop,	$1\frac{5}{8}$	192	"	6,790	$106\frac{1}{4}$
"	"	"	Modified Sawyer,	"	192	"	6,790	$106\frac{1}{4}$
"	"	"	"	"	192	"	6,790	$106\frac{1}{4}$
June,	"	Lyman Mills, Holyoke,	Whitin, alt'd to Saw'r Sp.,	$1\frac{7}{16}$	128	$4\frac{3}{4}$	5,597	117
"	"	"	"	"	128	"	5,597	117
"	"	"	"	"	128	"	5,597	117
"	"	"	"	"	128	"	4,633	135
"	"	"	"	"	128	"	4,633	135
"	"	"	"	"	128	"	4,633	135

SAWYER SPINDLE.

Draft.	Rov'g.	Yarn.	Ft. Lb. Frame.	Ft. Lb. Spin.	H. P. Frame	Spindle per H.P.	Remarks.
8	2.25	18 warp	1,002.67	4.478	1.823	123	Bob. av'ge ¼ full, { Hussey Banding. Bands tight.
8	2.25	18 "	968.75	4.32	1.761	127	" " " Bands adjusted.
8	2.25	18 "	1,159.42	5.176	2.18	106	" " " Single Banding.
8	2.25	18 "	1,113.63	4.97	2.03	110.6	" " " { Hussey Banding. 6.25 in. Trav., 1.8 oz Y'n.
7.84	1.62	12.33 "	643.23	5.025	1.179	110	" " " " " 2.3 "
7.84	1.62	12.33 "	696.5	5.48	1.275	100+	" " " Single Banding, " "
7.92	3.54	28 "	738.69	5.77	1.342	95.3	" " " " " Bands tight.
7.92	3.54	28 "	762.5	5.96	1.386	92.3	" " " " " "
8	2	16 "	718.39	4.99	1.306	110	" " " As running.
8	2	16 "	925.34	6.42	1.682	85.7	" " " " " Tight Belt.
8	1.54	12.30 "	732.1	5.72	1.296	96	" " " " " "
8	1.54	12.30 "	757.5	5.91	1.377	93	" " " " " "
9	1.85	16.65 "	964.28	4.64	1.753	118.6	" " " Ordinary Banding.
9	1.85	16.65 "	865.7	4.26	1.61	129	" " " " Hussey "
8.54	2.50	22 "	775.76	5.46	1.415	102	" " " As running. Bands tight.
8.54	2.50	22 "	804.35	5.58	1.462	98.5	" " " " " " "
7	3	21 "	567.16	4.43	1.031	124	" " " " " Bands easy.
....	....	......	335.82	2.63	....	....	" empty. Spindle only.
7.50	3	22.5 "	431.25	3.37	.784	163	" av'ge ¼ full. As running.
7.50	3	22.5 "	543.75	4.24	.988	130	" " " " "
8.80	1.65	14 "	1,207.93	5.39	2.144	102	" " " " "
8.14	3.68	32 "	858.95	4.47	1.561	123	" " " " "
...	....	....	556.72	2.9	....	....	Empty Bobbin. Spindle only.
7	4.35	30 "	980	5.57	1.792	99	Av'ge ¼ full Bobbin. As running.
7	4.35	30 "	1,010	5.74	1.836	97	" " " Frame just leveled.
6.89	3.64	25 "	1,166.66	6.63	2.141	83	" " "
7	1.90	13.69 "	1,229.16	6.4	2.235	} Av'ge 78.7	Empty Bobbin. } "Willimantic" Sawyer
7	1.90	13.69 "	1,456.25	7.58	2.648		Full " } Spindle.
10	2 dbl.	9 "	953.49	7.44	1.733	} Av'ge 72.2	Empty " { Regular Sawyer Spin.
10	2 "	9 "	1,000	7.81	1.818		Full " { Bands very tight.
9.5	2	19 weft	968.45	6.72	1.761	82	Bobbin ¼ full. New Frame. Had run 6 weeks.
9.5	2	19 "	959.2	6.66	1.744	82.6	" ¼ " Similar Frame.
9.5	2	19 "	849.5	5.9	1.545	93.2	" " " 1st Fr. repeated, slower sp'd.
8	2	16 "	989.58	6.87	1.8	80	" " " Sim. Fr. 4 weeks in opera'n.
7	2	14 "	908.16	6.3	1.651	87.3	" " " " 4 " "
7.35	1.50	11 "	561.22	3.9	1.02	140	" " " " 4 " "
8 24	2.61	21.50 wp.	1,347.06	7.01	2.44	....	Empty Bobbin.
8.24	2.61	21.50 "	1,617.65	8.42	2.941	....	Full "
8.24	2.61	21.50 "	1,482.35	7.717	2.695	71.3	Average "
....	....	23 weft	633.03	4.84	1.151	....	Empty "
....	....	23 "	633.03	4.84	1.151	113.6	Half full "
....	....	23 "	642.3	5.02	1.168	....	Full "
....	....	13 "	467.89	3.655	.851	} Av'ge 140	Empty " } Speed irregular all
....	....	13 "	477.06	3.726	.867		Half full " } through trials.
....	....	13 "	467.59	3.655	.851		Full "

RING SPINNING.—(*Continued.*)

Date.	Place.	Maker.	Diam. of Ring.	No. Spin.	Weight Spindle.	Rev. Spin.	Rev. F. Roll.
June, 1878	Lyman Mills, Holyoke,	Whitin, alt'd to Sawyer Sp.	$1\frac{7}{8}$ in.	128	$4\frac{5}{8}$ oz.	5,383	121
" "	" " "	"	"	128	"	5,383	121
" "	" " "	"	"	128	"	5,383	121
Oct., "	Pacific Mills,	"	$1\frac{5}{8}$	160	$3\frac{3}{4}$	5,851	70.5
" "	" "	"	"	160	"	5,851	70.5
" "	" "	"	"	160	"	5,851	70.5
" "	" "	"	"	160	"	5,893	71
" "	" "	"	"	160	"	5,893	71
" "	" "	"	"	160	"	5,893	71
" "	" "	"	"	160	"	5,893	71
" "	" "	"	"	160	"	5,893	71
" "	" "	"	"	160	"	5,893	71

PEARL

Date.	Place.	Maker.	Diam. of Ring.	No. Spin.	Weight Spindle.	Rev. Spin.	Rev. F. Roll.
June, 1874	Lawrence Manu. Co.,	Lowell Ma. Shop	$1\frac{5}{8}$ in.	224	$6\frac{1}{2}$	6,150	99
July, 1875	Pacific Mills, Lawrence	Davis & Furber, No. 3 Sp.	$1\frac{1}{2}$	160	$4\frac{15}{32}$	5,800	62
" "	" " "	"	"	160	"	5,800	70
Aug., "	Stark Mills,	Davis & Furber, No. 1 Sp.	2	144	$6\frac{3}{4}$	5,300	106
" "	" "	"	"	144	"	5,470	108
Oct., "	Manchester Mills,	Saco W. P. Co., No. 2 Sp.	$1\frac{5}{8}$	192	$6\frac{1}{4}$	5,830	71
Dec., "	Jackson Co.,	Lowell Shop,	$1\frac{9}{16}$	208	6	4,340	107
Jan., 1876	Salmon Falls,	Davis & Furber,	$1\frac{5}{8}$	144	$4\frac{1}{8}$	5,500	63
" "	" "	Pearl Bobbin Saco Spindle Cut-off Tip	"	144	9	5,800	74
April, "	York Co., Saco,	Saco W.P.Co. Throstle alt'd	"	128	$6\frac{1}{4}$	6,140	$80\frac{1}{2}$
" "	" " "	"	"	128	"	5,700	82
" "	" " "	Lowell M. Sh. Throstle alt'd	"	128	$5\frac{1}{2}$	5,515	$82\frac{1}{2}$
" "	" " "	"	"	128	"	5,870	84
Nov., 1877	Atlantic Mills,	Davis & Furber, Spind. in Lowell Fr'me	"	176	$8\frac{3}{4}$	5,022	88
" "	" "	Lowell Ma. Shop	"	176	$4\frac{1}{8}$	6,059	73
" "	" "	"	"	176	"	6,059	73
" "	Jackson Co., Nashua,	"	$1\frac{3}{4}$	192	$4\frac{3}{4}$	5,564	115.5
" "	" " "	"	"	192	"	5,564	115.5
Sept., 1878	China Mill,	Whitin, alt'd to P'l Sp. No. 2	$1\frac{5}{8}$	128	"	7,400	77
" "	" "	"	"	128	"	7,400	77
" "	" "	"	"	128	"	7,400	77
Oct., "	Massachusetts Mill, Lowell,	Lowell M. Sh. P'rl Sp. No. 1	$1\frac{3}{4}$	160	5	5,080	107
" "	"	"	"	160	"	5,080	107
" "	"	"	"	160	"	5,080	107
" "	Pacific Mills, Lawrence,	Davis & Furb. P'rl Sp. No. 3	$1\frac{1}{2}$	160	$4\frac{15}{32}$	6,006	66
" "	"	"	"	160	"	6,006	66
" "	"	"	"	160	"	6,006	66
" "	"	"	"	160	"	6,006	66
" "	"	"	"	160	"	6,006	66
" "	"	"	"	160	"	6,006	66

SAWYER SPINDLE.—(Continued.)

Draft	Rov'g	Yarn	Ft. Lb. Frame	Ft. Lb. Spin	H. P. Fr'me	Sp. per H. P.	REMARKS	
.....	.....	12 warp	800	6.25	1.355	} Av'ge { 80—	Empty Bobbin	} Speed irregular all through trials.
.....	.....	12 "	911.11	7.12	1.056		Half full "	
.....	.....	12 "	951.11	7.43	1.729		Full "	
6.79	4.40	28.50"	829.54	5.18	1.508	......	Empty Bobbin	} Pearl Bobbin, weight 240.6 grains, yarn on Bobbin 458.4 grains.
6.79	4.40	28.50"	852.27	5.326	1.550	......	Full "	
6.79	4.40	28.50"	840.40	5.25	1.529	104.5	Average "	
6.79	4.40	28.50"	832.91	5.206	1.514	......	Empty Bobbin	} Hybrid Bobbin, between Pearl and Sawyer, w'ght 1 obbin 213.3 gr., weight Yarn on Bob. 470.3 gr
6.79	4.40	28.50"	843.84	5.305	1.543	......	Full "	
6.79	4.40	28.50"	840.88	5.256	1.529	104.5	Average "	
6.79	4.40	28.50"	819.77	5.12	1.490	......	Empty Bobbin	} Sawyer Bubbin, weight 295.3 grains, Yarn on Bobbin 465 grains.
6.79	4.40	28.50"	883.72	5.52	1.607	......	Full "	
6.79	4.40	28.50"	851.74	5.32	1.548	103.5	Average "	

SPINDLE.

8	2.25	18 "	1,411.11	6.299	2.565	87.5	Average ⅓ full.	Hussey banding, tight.
8.14	4.40	38 "	629.95	3.94	1.146	140	" "	Ordinary banding.
6.94	4.40	28 "	648.33	4.05	1.161	136	" "	" "
8	1.55	12.30"	1,152.50	7.97	2.095	69	" "	} Extra large Bobbin. 2,000 yards yarn. New. Higher speed.
8	1.55	12.30"	1,162.5	8.07	2.113	68	" "	
7.66	4	30 "	980	5.10	1.782	108	" "	
9	1.50	13.50w'ft	1,108	5.33	2.014	103.24	" "	
7.07	4.29	29 warp	955	6.63	1.736	83	" "	Bands very tight.
8.54	2.50	22 "	865.67	6.01	1.574	91½	" " "	tight.
7	3	21 "	716.99	5.61	1.313	98	" " "	in good order.
7	3	21 "	611.11	4.77	1.111	115	" " "	" "
7	3	21 "	594.70	4.65	1.082	118	" " "	" "
7	3	21 "	644.96	5.04	1.173	109	" " "	" "
7.60	2	15 "	824.07	4.685	1.500	117	Bobbin ⅓ full.	{ Pearl Spindle, like original Model in Patent Office.
7.60	3.68	28 "	868.85	4.937	....	......	" empty.	Chambered tip, P'rl Sp. No. 2
7.60	3.68	28 "	942.60	5.356	....	......	" "	Plugged " " "
7	1.90	13.69w'ft	1,364.58	7.10	2.481	} Av'ge { 70.7	" "	} Pearl Spin. No. 1. Frame New. Bands and Belts tight.
7	1.90	13.69 "	1,625	8.46	2.954		" full,	
7.72	3.21	28.50wp.	956.21	7.47	1.739	......	" empty,	} Rolls too slow for speed of Spindle.
7.72	3.21	28.50 "	1,138.46	8.89	2.070	......	" full,	
7.72	3.21	28.50 "	1,047.3	8.18	1.904	67+	" average.	
.....	.....	13 warp	812.5	5.08	1.477	......	Empty Bobbin.	}
.....	.....	13 "	914.44	5.90	1.717	......	Full "	
.....	.....	13 "	878.47	5.49	1.597	100+	Average "	
8.14	4.40	38 "	736.21	4.60	1.339	......	Empty Bobbin,	} Regular Pearl Bobbin, 241 grains. Yarn on Bobbin 421 gr.
8.14	4.40	38 "	762.07	4.76	1.385	......	Full "	
8.14	4.40	38 "	749.14	4.68	1.362	117.75	Average "	
8.14	4.40	38 "	736.21	4.60	1.339	......	Empty "	} Lower Bush of Bobbin reamed out. Bobbin, 2'0·6 grains. Yarn on Bob. 421.1 grs.
8.14	4.40	38 "	753.45	4.70	1.370	......	Full "	
8.14	4.40	38 "	744.83	4.65	1.355	118.25	Average "	

RING SPINNING.—(*Continued.*)

DATE.	PLACE.	MAKER.	Diam. Ring.	No. Spin.	Wt. Spin.	Rev. Spindle.	Rev F. Roll.	Draft.
July, 1875	Augusta, Me.,	Fales & Jenks,	1$\frac{3}{4}$ in.	128	7$\frac{3}{4}$ oz.	5,200	63	6.98
" "	Pacific Mills,	" "	1$\frac{1}{2}$	160	3$\frac{3}{4}$	6,200	58	4.6
" "	" "	" "	"	160	"	6,200	58	4.6
Dec., 1877	Whittenton Mills,	Whitin, alt'd to Rabbeth,	1$\frac{13}{16}$	128	4	5,135	130	10
BIRKENHEAD								
Aug., 1875	Stark Mills,	C. Lamphear,	2	132	5$\frac{1}{4}$	5,460	108	8
Dec., 1877	Whittenton Mills,	Whitin, altered,	1$\frac{13}{16}$	128	4	4,600	128	10
EXCELSIOR								
April, 1875	Stark Mills, Man'r, N. H.,	Bridesburg Co.,	1$\frac{3}{4}$	204	6	4,730	104	8
" "	"	"	1$\frac{9}{16}$	204	6	4,520	110	8.24
Aug., "	"	Bridesburg Co., Cupholder on Spin.	2	204	8$\frac{1}{2}$	4,785	100	8
Oct., "	"	Excelsior Spin.,	1$\frac{9}{16}$	204	6	4,446	110	8.3
" "	"	"	1$\frac{3}{4}$	204	6	4,660	104	7.84
Nov., "	"	" repeat'd	2	204	6	4,660	105	7.84
" "	"	Cupholder Sp. rep.	2	204	8$\frac{1}{2}$	4,930	102	8
LOWELL MACHINE SHOP								
Aug., 1876	Boott Mills,	Lowell Shop,	1$\frac{5}{8}$	224	5$\frac{1}{2}$	6,480	103	8.80
PERRY								
Aug., 1875	Stark Mills,	Amoskeag Shop,	2	128	2$\frac{5}{8}$ fl'r	5,588	110	8
MASON'S NEW SINGLE								
Dec., 1877	Whittenton Mills, Taunton,	Wm. Mason, New Sgl. Rail Sp. set in Jaws,	2	128	2$\frac{5}{8}$	5,135	130	10
" "	"	"	2	128	2$\frac{5}{8}$	5,135	130	10
AMOSKEAG CO.'S								
Jan., 1878	Namaske M., Manchester,	Amosk. Co., Short Str't Clutch Spin.	2$\frac{1}{4}$	120	6	3,955	100	5.5
" "	"	"	2$\frac{1}{4}$	120	6	3,955	100	5.5
" "	"	"	2$\frac{1}{4}$	120	6	4,520	115	6.4

RABBETH SPINDLE.

Roving.	Yarn.	Ft. Lb. Frame.	Ft. Lb. Spin.	H. P. Frame	Spindle per H. P.	REMARKS.
4.4	20.50 wp.	916.66	7.16	1.666	77	Bobbin half full. Old Pattern Spindle.
11.5	45 "	632.14	3.95	1.149	} Av'ge { 132	" empty. } Average, 4.17 lb. per Spin.
11.5	45 "	703.59	4.4	1.279		" full.
1.8 dbl.	9 "	953.49	7.45	1.726	73.8	" half full. Bands tight.

SPINDLE.

1.54	12.33 "	1,082.03	8.19	1.963	67	"	" Extra large, 2,000 yds. of Y'n
1.8 dbl.	9 "	1,038.37	8.11	1.888	67.8	"	" [when full.

SPINDLE.

1.52	12.30 "	1,087.31	5.33	1.977	103	"	" As running.
1.62	13.5 wft.	1,136.36	5.56	2.065	100—	"	" " "
1.54	12.3 wp.	1,506.16	7.36	2.748	74.5	"	" New Frame. As running. [Large Bobbin.
1.6	13.5 wft.	931.25	4.56	1.693	120	"	" As running.
1.6	12.3 wp.	1,261.72	6.19	2.292	89	"	" Bands very tight. As run'g.
1.6	12.3 "	890.63	4.36	1.619	126	"	" Bands eased.
1.52	12.3 "	1,332.03	6.54	2.422	84	"	" Same Frame tested in Aug.

COMBINATION SPINDLE.

1.65	14 wp.	1,349.2	6.02	2.453	91¾	"	" Comb'n Pearl and Sawyer Spin.

SPINDLE.

1.54	12.3 wp.	1,012.04	7.91	1.84	70	"	" New Frame, large Bobbin.

RAIL SPINDLE.

1.8 dbl.	9	wp.	966.29	7.55	1.757	} Av'ge 64.6	Empty Bobbin. Bands tight.
1.8 "	9	"	1,213.36	9.48	2.206		Full " " "

SHORT SPINDLE.

1	5.5 wp.	691.86	5.77	1.258	} Av'ge 87.3	Bobbin empty. } Stock all Waste for
1	5.5 "	819.87	6.88	1.49		" full. Coarse Duck.
1.25	7 "	1,069.77	8.75	1.945	63	" half full. Cotton Stock for Duck Wp.

MULE SPINNING.

Date.	Place.	Description.	No. Spin.	Rev. Spin.	Length Stretch.	Time Stretch.
Mar., 1874	Eagle Mill, Taunton,	W. Rouse, by Dean Cotton Machine Co.,	592	4,876	60 in.	17 sec.
" "	" "	"	592	4,896	60	17
" 1875	China Mill, Suncook,	Wm. Mason's,	512	3,889	60	19⅜
" "	Webster Mill, Suncook,	"	512	3,540	60	21
" "	Pembroke Mill, Suncook,	"	512	3,780	60	20
" "	Newton M., New'n, Mass.	"	384	3,780	60	21
" "	" "	"	384	3,780	60	21
April, 1877	Lockwood Mills, Waterville, Me.,	Saco W.P.Co., "Parr Curtis," Long Spin.,	640	4,360	64	18½

COTTON TWISTERS.

Date.	Place.	Description.	Diam. Ring.	No. Sp.	Rev. Spin.	Ft. Lb. per Sp.	H. P.
Feb. 1875	Renfrew Mfg. Co.,	Collins,	2 in.	132	3,452	7.57	1.818
"	South Adams, Mass.,	"	2½	100	3,700	10.47	1.902
"	" " "	Same, altered to Sawyer Spindle.	"	100	3,700	8.82	1.603
Jan. 1877	Conant Thread Co., Pawtucket, R. I., No. 4 Mill,	Fales & Jenks, No. 80 Yarn,	1½	150	5,066	8.8	2.4
"	"	Rabbeth Sp., No.80 Y'n,	"	150	5,066	8.47	2.309
"	"	1st Twist, No. 100	"	"	150	5,066	8.8	2.4
"	"	" " " 120	"	"	150	5,066	8.67	2.364
"	"	2d " " 80	"	1⅖	150	4,560	10.77	2.936
"	"	" " " 100	"	"	150	4,560	9.8	2.673
"	"	" " " 120	"	"	150	4,560	9.73	2.654
"	Pawtucket, R. I., No. 2 Mill,	1st " " 80	"	1½	150	5,428	8.58	2.341
"	"	" " " 80	"	"	150	5,428	7.83	2.136
"	"	" " " 32	"	"	150	5,428	9.5	2.591
"	"	2d " " 32	"	1⅘	150	4,880	11.29	3.079
"	"	" " " 80	"	"	150	4,657	8.69	2.37
"	"	" " " 80	"	"	150	4,657	8.89	2.424

FLAX MACHINERY.

Date.	Place.	Description.	Delivery Rolls.	Ft. Lbs.	Horse-Power.
Apr., 1874	Stark Mills,	1 Tow Card, 6 Workers,	165 Rev.	1,807.40	3.286

MULE SPINNING.

Draft.	Roving.	Yarn.	START.		DRAFT.		TWIST & BACK.		Average Ft. Lb.	Lb. per Spin.	Average H. P.	Spindles per H. P.
			Sec.	Ft. Lb.	Sec.	Ft. Lb.	Sec.	Ft. Lb.				
8	3.6	28.50	3	2,252	9	1,586	5	633	1,423	2.59	2.4	229*
8	3.6	28.50	3	1,905	9	1,429	5	574	1,260	2.29	2.13	258†
9.16	4.08	37	3	1,716.66	12	1,383.31	4⅔	384.33	1,044.17	1.898	2.04	270
9.16	4.08	37	2	1,448	12	1,055	7	323	848.52	1.548	1.64	332
9.4	3.91	37	2	1,610	11	1,310	7	510	1,060	1.927	2.07	266
9.16	4	36	..	....	15	866.6	6	341.6	716.6	1.303	1.85	294 (old)
9.16	4	36	2	1,333.3	13	1,000	6	416.6	865	1.573	2.25	244 (new)
8.5	4.5	36	5	2,635.13	8½	1,959.46	5	878.35	1,850	3.364	2.89	190

COTTON SPOOLERS.

Date.	Place.	Maker.	No. Spin.	Rev. Spindle.	Ft. Lb. per Sp.	Horse- Power.	No. Yarn.
Mar., 1875	China Mills,	Saco W. P. Shop,	80	775	1.48	.23	29
"	Webster Mills,	"	80	750	1.81	.264	29
"	Pembroke Mills.	"	80	600	1.68	.244	29

WARPERS.

Date.	Place.	Maker.	No. Yarn.	No. Thr'ds.	Speed.	H. P.
Mar., 1875	China Mills,	Saco W. P. Shop, "Iron Tip" Spools,	28	438	40 yds per min.	.185
"	"	Saco W. P. Shop, "Skewers,"	28	438	40 " "	.142
"	Webster Mills,	"	28	292	48 " "	.140

SLASHERS.

Mar., 1875	China Mills,	Howard & Bullough,	28	1,752	25 yds. per min.	1.242
"	Webster Mills,	"	28	1,752	32 " "	1.694
Aug., 1876	Boott Mills,	Lowell Shop, "Hot Air,"	14	1,300	27 " "	1.63

* Wet day.   † Fair day.

COTTON LOOMS.

Date.	Place.	Maker.	Width.	No. Warp.
May, 1874	Stark Mills,	Amoskeag, Bag Loom,	24 in.	10 dbl.
"	"	" Crash "	24	.....
"	"	Thomas, Diaper "	24	.....
Nov., 1874	Manchester Mills,	Bridesburg Co.,Clip'r L'm,*New*	30	70
"	Stark Mills,	Amoskeag Co.,	37	12.30
Jan., 1875	Arlington Mill,	S. T. Thomas,	30	70
"	Methuen, Alpacas, &c.,	Hodson, English,	30	70
"	"	Bridesburg, Clipper,	30	60
"	"	Same, tested singly,	30	60
Mar., 1875	Webster Mill, Suncook, N. H.,	Mason's,	30	28
"	"	"	36	28
"	Newton Falls, Newton, Mass.,	"	30	28
"	"	Empire Loom,	30	28
April, 1875	Manchester Mills,	Clipper "	30	70
"	"	Bridesburg Co., Alpacas,	30	70
"	"	"	30	70
April, 1877	Lockwood M., Waterville, Me.,	Lewiston M. Sh., New Looms,	40	25
"	"	"	30	30

WOOLEN MACHINERY.

CARDS.

Date.	Place.	Maker & Description.	Width.	Diam	Rev.	Lb. per Day.	Ft. Lbs.	H. P. each.
Aug., 1876	Saxonville, Mass.,	Davis & Furber, 1st Breaker,	48 in	48 in	108	225	484.30	.88
"	"	Davis & Furber, 2d Breaker,	48	48	108	225	426.47	.775
"	"	Davis & Furber, 2d Finisher,	48	48	108	225	423.91	.771
"	"	1 W'st'd Card & Comb	40	44	115	300	1,162.28	2.113
"	"	"	34	36	125	300	952.28	1.732
Jan., 1877	Passaic, N. J.,	1 Wool Card with Oiler Breaker,	60	48	84	300	667.27	1.231
"	Rittenhouse Co.,	1 Finisher with 4 Rubbers,	60	40	87	300	667.27	1.231

PICKERS.

Date.	Place.	Maker & Description.	Width.	Diam	Rev.	Lb. per Day.	Ft. Lbs.	H. P. each.
Aug., 1876	Saxonville Mills,	1 Wool Picker,	36 in	36 in	360	2000	3,161.11	5.747
"	"	1 " "	36	36	480	2500	3,640	6.618
"	"	1 Rag "	24	36	560	1200	4,028.57	7.324
Jan., 1877	Rittenhouse Co.,	1 " "	24	36	770	1500	4,861.54	8.839
"	Passaic, N. J.,	1 Waste Lumper,	48	16	810	....	743.24	1.351
"	"	1 Parkhurst Burr M.	36	....	353	....	2,341.18	4.258

COTTON LOOMS.

No. Weft.	Picks per in.	Picks per Min	Ft. Lbs. Loom.	H. P. Loom.	No. test'd	Total H. P.	Looms p. H.P.	Notes.
8 double	......	126	214.28	.39	6	2.338	2.56	Counter shaft for 16 Looms, .881 H.P.
....	......	136	89.74	.163	6	.979	6.13	" " 8 " .22 H.P.
....	......	84	82.91	.151	4	.603	6.62	" " 6 " .145 H.P.
No. 36 w'st'd	43 × 80	188	235.4	.428	1	......	2.32	Taken in quantity would be probably
13.56	48 × 48	158	180.5	.328	7	2.297	3.05	[3.5 Looms per H.P. Counter shaft for 8 Looms, .53 H.P.
32	43 × 80	142	80	.145	8	1.164	7	" " 8 " .282 H.P.
....	......	142	107.5	.195	2	.391	5+	
32	43 × 80	172	110.9	.202	8	1.614	5	" " 8 " .25 H.P.
32	43 × 80	172	148	.27	1	......	3.38	Single Loom, tests doubtful. (?)
36	64 × 64	154	60	.109	10	1.091	9+	{ 200 ft. Shafting, average, 2¼ Diam.
36	64 × 64	154	74.54	.135	10	1.355	7.4	{ 208 Rev. per min. 1.487 H.P.
36	64 × 64	140	75.2	.136	13	1.758	7.4	Counter shaft. .833 H.P.
36	64 × 64	175	85.06	.155	4	.619	6.45	" " .088 H.P.
36	43 × 80	140	143	.26	7	1.818	3.84	3 Short Counters, 230 Rev. .545 H.P.
36	43 × 80	154	170.83	.31	6	1.863	3.22	3 " " 250 " .568 H.P.
36	43 × 80	180	223.21	.406	7	2.861	2.46	3 " " 292 " .613 H.P.
29.5	68 × 72	148	121.49	.219	4	.876	4.57	2 " " 240 " .096 H.P.
32	64 × 64	160	103.66	.188	3	.565	5.32	2 " " 240 " .086 H.P.

WOOLEN MACHINERY.—(*Continued.*)

SPINNING.

Date.	Place.	Description.	No. Sp.	Ft. Lb. Stretch.	Ft. Lb. Twist.	Ft. Lb. Back.	Ft. Lb. Av'ge.	H P. Av'ge.
Aug., 1876	Saxonville Mills,	1 Woolen Mule,	150	704.54	568.18	318.18	536.91	.976
Jan., 1877	Rittenhouse Co.,	1 " "	200	612.33	546.68	380	534.66	.972
"	"	1 " "	364	972	892	612	854.38	1.553

LOOMS.

Date.	Place.	Description.	Width Loom.	No. Harn's.	Picks per Min	Ft. Lb. per Sec.	H. P.	Looms per H.P.
Aug., 1876	Saxonville, Mass.,	Blanket Loom,	90 in	4	68	101.85	.185	5.4
Jan., 1877	{ Rittenh'seCo., } { Passaic, N. J., }	"	90	4	68	202.08	.387	2.58

FINISHING MACHINERY.

Date.	Place.	Description.	Diam	Rev.	Ft. Lb.	H. P.
Jan., 1877	Rittenhouse Co.,	1 Hydro Extractor,	60 in	375	2,340	4.255
"	"	1 Rotary Fulling Mill, Double,	16	96	1,676.47	3.081
"	"	1 Rinser to above,	16	94	1,504	2.844
"	"	1 Broad Gig, 108 in.,	48	148	833.33	1.515

WORSTED MACHINERY.

Date.	Place.	Description.	Diam Rolls.	Rev. Rolls.
Oct., 1871	Manchester Mills,	Eng. Spin. Frame, "Flier,"	4 in.	24
"	"	" alt'd to "Cap Spin.,"	2¾	44
Dec., 1871	"	" Platt Bros. "Flier,"	4	22
Jan., 1873	"	" " "	4	20
"	"	" alt'd to "Cap Spin.,"	4	42
Nov., 1874	"	" " "	4	48
Jan., 1875	{ Arlington Mills, Methuen, Mass., }	1 English Flier Frame,	4	23.5
"	"	1 " "	4	23.5
"	"	1 " " alt'd to Cap,	4	41
"	"	1 2-Can Gill Box, 1st Proc.,	....	102 bars per min.
"	"	1 2-Can " 2d "	....	156 " "
"	"	1 2-Spin. " 3d "	....	146
"	"	1 4-Spin. 1st Rov.,4th "	....	88
"	"	1 6-Spin. 2d " 5th "	....	92
"	"	1 6-Spin. 3d " 6th "	....	48
"	"	1 Dandy Roving, 7th "	....	20.5
"	"	1 1st Preparing Gill,	....	62 bars per min.
"	"	1 2d " "	....	82 " "
"	"	1 3d " "	....	80 " "
"	"	1 4th " "	....	90 " "
"	"	1 5th " "	....	130 " "
"	"	1 Lister Comber,	....	36 sweeps "
"	"	1 Noble "	....	293 rev. sh. "
Aug., 1876	Saxonville Mills,	1 Single Bobbin, 1st Rov'g,	....	....
"	"	1 Double " 2d "	....	....
"	"	1 Dandy or 3d "	....	....
"	"	1 Flier Spinning Frame,	4	40
"	"	1 Twister,	....	....
Mar., 1878	{ A. T. Stewart & Co., Worsted Mills, Glenham, N. Y., }	1 Wool Washer,	14	7
"	"	1 Yarn Squeezer,	14	5
"	"	1 1st Preparing Gill,	....	....
"	"	1 Screw Gill Balling Head,	....	....
"	"	1 Noble Comber,	....	....
"	"	1 1st Screw Gill Drawing,	....	....
"	"	1 Screw Gill Slubber,	....	....
"	"	1 " 1st Roving,	....	....
"	"	1 Dandy Roving,	....	....
"	"	1 1st Bal'g Box, Sliv. fr'm Cd.	....	....
"	"	1 Throstle Spinning Frame,	4	48
"	"	1 " " "	4	16
"	"	1 Twister,	....	....
"	"	1 Redoubler,	....	....
"	"	1 Reel,	....	....
"	"	1 Print Yarn Winder,	....	....
"	"	1 Warp Beamer,	....	shaft, 337
"	"	1 Linen Skein Winder,	....	....
"	"	1 Warp Dresser,	....	....
"	"	1 Small Skein Spooler,	....	....
"	"	1 Quill Winder,	....	....
"	"	1 27-ft. Print Drum, 3 ft. wide	....	....
"	"	1 18-ft. 9-in. " 3 " "		

WORSTED MACHINERY.

No. Sp.	Rev. Spindles.	Draft.	Roving.	Yarn.	Ft. Lb. Machine.	Ft. Lb. per Spin.	H. P. Machine.	Spindle per H.P.
120	2,474	8.8	3.3 gr.	No. 34	810	6.75	1.473	81.5
120	5,025	8	2.05	52	968	8.07	1.759	68
132	2,500	8.8	3.3	34	744.5	5.64	1.354	97.5
144	2,470	8.8	3.3	34	1,170	8.12	2.127	67
144	5,025	8.8	3.3	34	1,618	11.23	2.941	49
144	5,820	8.8	3.3	34	2,276	15.1	4.135	36.4
144	2,356	8.9	3.1	32	962	6.68	1.741	82.3
128	2,356	8.9	3	30	886.5	6.92	1.612	80
144	4,575	8.9	3	30	1,440	10	2.618	55
......	....	....	....	....	625	....	1.126	....
......	....	....	....	....	185	....	.336	....
2	62	....	....	....	170	....	.309	....
4	148	....	....	....	560	....	1.018	....
6	200	....	....	....	510	....	.942	....
6	240	....	....	....	210	....	.382	....
30	655	9.9	....	....	517.4	17.27	.94	31.8
......	....	....	....	....	487.5	....	.886	....
......	....	....	....	....	404	....	.735	....
......	....	....	....	....	321	....	.583	....
......	....	....	....	....	466	....	.848	....
......	....	....	....	....	345	....	.627	....
......	....	....	....	....	813	....	1.478	....
......	....	....	....	....	748	....	1.361	....
1	150	....	....	....	201.18	....	.366	....
2	200	....	....	....	241.66	....	.439	....
12	250	....	....	....	307.5	25.66	.559	21.4
128	3,110	....	....	....	1,264.29	9.88	2.3	55.6
108	2,650	....	....	....	1,600	14.81	2.91	37
......	....	Rakes,	14 sw'ps	per min.	1,066.82	....	1.939	....
......	....	....	....	....	345.33	....	.628	....
......	100 bars per min.	....	....	....	568.62	....	1.033	....
......	205 " "	....	....	....	225 45	....	.41	....
......	240 rev. sh. "	....	....	....	866.66	....	1.573	....
4	250 " "	....	....	....	289.56	....	.526	....
8	265 " "	....	....	....	694.12	....	1.262	....
24	265 " "	....	....	....	694.12	....	1.262	....
24	833	....	....	....	734.44	30.6	1.335	....
......	250 rev. sh. "	....	....	No.	85	....	.154	....
144	2,289	....	....	No. 5	1,407.14	9.77	2.558	...
144	2,289	....	....	12	1,164.29	8.08	2.117	....
124	1,526	....	....	....	1,028.37	8.3	1.87	....
42	....	....	....	....	148.4	....	.27	....
41 sk	25	....	....	....	107	....	.194	....
20 dr ums	150	....	....	....	373.33	....	.679	....
......	....	....	....	....	165	....	.3	....
25 dr ums	198	....	....	....	336	....	.611	....
......	....	....	....	....	438.2	....	.797	....
......	....	....	....	....	168.18	....	.306	....
30 dr ums	1,234	....	....	....	353.31	....	.642	....
......	....	....	....	....	114	....	.207	....
......	....	....	....	....	121.11	....	.22	....

WORSTED MACHINERY.—(Continued.)

Date.	Place.	Description.	Picks per Min.	Ft. Lb.	H. P.
Mar., 1878	A. T. Stewart & Co., Worsted Mills, Glenham, N. Y.,	1 18-ft. 9-in. Pr. Dr., 6 ft. wide,	..	122.55	.223
"	"	1 ¼ Tapestry Brussels L'm,	60	390.74	.71
"	"	1 ¼ " " "	80	301.85	.549
"	"	1 ¼ Dandy " "	80	464.29	.844
"	"	1 ¼ 5 Frame Jacquard "	60	573.69	1.043
"	"	1 Set 7 Dry Cans, 24 in. Diam.,	..	46.33	.084
"	"	1 Carpet Shear,	..	706.66	1.284
"	"	1 Winding & Meas'g Mach.,	..	200	.364

MISCELLANEOUS MACHINERY.

Date.	Place.	Description.	Rev.	Ft. Lb.	H. P.
May, 1874	Stark Mills,	1 Fan in No. 3 Pick'r, 3 ft. Di.,	827	1,396.55	2.54
Mar., 1875	Crosby, Morse & Co., Boston,	1 Diamond Grinding Machine, 12 in. Diam.,	2100	364.03	.662
July, 1875	P. C. Cheney & Co., Goffstown, N. H.,	1 Pulp Grinder, 4 ft. Di., 4 in. Face, running alone,	425	666.66	1.212
"	"	Same with 1 Sliding Box up, 100 lbs. pressure,	425	1,633.33	2.97
"	"	Same with 2 Boxes up, opposite sides of stone,	425	4,466.66	8.121
"	"	Same with 3 Boxes up, opposite sides of stone,	425	6,873.24	12.5
"	"	Same with 4 Boxes up, opposite sides of stone, Belt slipped on this trial,	380	7,859.15	14.29
Oct., 1875	Manchester Mills, Manchester, N. H.,	1 Cloth Shear, Print Cloth, 4 Cutters,	2500	1,733.33	3.151
"	"	1 Cl. sh., 36 in. Cl., 4 Cutters,	2500	2,566.66	4.666
May, 1876	Hopedale, Mass., Geo. Draper & Sons	1 Soda Pump in Shop, 4 in. x 8 in.,	18	94.8	.172
"	"	1 Ring Polish'g Lathe, empty,	....	244.83	.445
"	"	Same in Full Work,	....	451.73	.821
"	"	2 Mill'g Mach., 2 Rings each,	....	555.15	1.009
"	"	1 Small Engine Lathe,	....	44.12	.08
"	"	"Duster" in Foun. (Rattlebox),	....	416.66	.755
"	"	" " " Smaller,	....	313	.569
Dec., 1876	Douglass Braid M., Providence, R. I.,	1 Bench, 32 Braiders,	....	216.15	.393
"	"	Same, 22 "	....	177.88	.323
"	"	Difference, 10 "	....	38.27	.07
"	"	1 Bench, 64 "	....	338.62	.616
"	"	1 Skein Spooler, 100 sk., including Counter Shaft,	....	147.83	.269

MISCELLANEOUS MACHINERY.—(Continued.)

Date.	Place.	Description.	Rev.	Ft. Lb.	H.P.
Dec., 1876	Douglass Braid Mill., Prov.,	1 Tagging Machine,	....	39.13	.071
"	"	1 Reel Bench, 10 Reels,	....	25.14	.046
"	"	1 Braid Finishing Mach.,	....	245.63	.447
"	"	1 Straightening "	....	61.29	.111
"	"	1 Balling Head,	....	12.5	.023
"	Adams & Shaw, Prov., Silverw'e.	1 Sp'd Lathe, turn'g Cups,	400	130	.236
"	"	1 " " Wood,	1300	80.43	.146
"	"	1 Chuck " Dies,	150	53.84	.098
"	"	1 Polish'g Buffer, 6 in. Di.,	2950	360	.654
"	"	1 " Spin., 1 "	2800	280	.509
"	"	1 Emery Wheel,12 "	1200	187.5	.341
"	Robinson's Sh., Jewelry, Prov.,	1 pr. Flat'ng R'ls for Wire	....	113.33	.206
"	"	1 " break'g d'n R'ls for Pl.	....	977	1.773

SUMMARY OF POWER OF MILLS.

Mill A.—*Heavy Sheetings. Average No. of Yarn spun, 12.75.*

Description.	Total Spin.	Speed.	H.P. ca.	H. P.
1 Stick Whipper,	....	....	....	1
1 Double Creighton Willow,	....	....	....	12
2 Bacon Willows,	....	....	....	4
3 Kitson, 2 Beaters, 1st Pickers,	....	Beat'rs, 1,500 Rev.	5	15
4 " 2 " 2d "	....	" 1,500 "	5	20
3 Whitins, 3 " "	....	" 2,000 "	4.5	13.5
3 Amoskeag, 2 " "	....	" 1,500 "	4.5	13.5
96 36-in. Breaker Cards,	....	Cylin'r, 126 "	.144	13.824
2 Lap Heads to same,	....	....	1.25	2.5
112 Finisher Cards,	....	" 126 "	.144	16.128
16 Railway Heads to same,	....	....	.687	10.992
8 Frames, 1st Drawing, 8 Deliv's each,	....	per Delivery,	.11	7.04
8 " 2d " 8 "	....	"	.11	7.04
16 Lowell Speeders, 30 Spindles each,	480	Spindle, 720 Rev.	.845	13.52
28 " " 52 "	1,456	" 904 "	1.087	30.436
163 " Throstle Spin. Fr., 128 Spin. each,	20,564	" 4,100 "	1.7	277 1
5 Sawyer Spin. Ring " 128 "	640	" 5,700 "	1.3	6.5
2 Old St. Lowell " " 144 "	288	" 4,500 "	1.5	3
2 Platt Bros' Mules, 624 "	1,248	....	3	6
4 3¼-in. Ring Twisters, 80 "	320	" 2,812 "	1.467	5 8CS
5 Spoolers, 50 "	250	" 600 "	.25	1.25
11 " 40 "	4,400	" 600 "	.2	2.2
6 Winders for Filling, 100 "	600	" 2,900 "	1.5	9
6 " " 50 "	300	" 2,900 "	.75	4.5
2 Manchester Warpers,	·	....	.171	.342
10 English "	....	....	.125	1.25
2 Slashers,	....	....	1.5	3
677 36-in. Looms,	....	125 Picks	.16	108.32
1 Banding Machine,	....	....	....	.5
Total Machinery =	....	....	....	609.31
Very Heavy Shafting, estimated at 15% =	....	....	....	91.396
H. P. per 1,000 Sp.=30.41=33 Sp. per H.P.				
Total=	....	....	....	700.706

SUMMARY OF POWER OF MILLS.—(Continued.)

MILL A.—(Continued.)

Per cent. of Power of Machinery,
{ Picking and Carding, 29.62
Spinning, 48.02
Dressing, 4.58
Weaving, 17.78
——
100 }

MILL B.—Denims, Ticks, etc. Average No. of Yarn, 11.

DESCRIPTION.	Total Sp.	Speed.	H.P. ea.	H. P.
2 Kitson Comp'nd Opener H'ds, 2 Beat'rs,	.....	Beat'rs, 1,100 Rev.	3	6
5 " " " " 3 "	.....	1,100 "	5.5	27.5
5 Amoskeag 1st Pickers, 2 "	.....	1,100 "	4.86	24.3
10 " 2d " 2 "	.....	" 1,026 "	3.048	30.48
264 Breaker Cards,	.....	Cylin'rs, 110 "	.144	38.016
24 Railway Heads,	.....	F. Roll, 400 "	.687	16.488
8 Frames, 4 Roll Drawing, 8 Deliv's each,	.....	" 240 "	del .072	9.168
8 " 4 " 2d " 8 "	.....	" 240 "	" .074	9.456
8 Lowell Speeders, 6 in. × 12 in., 30 Sp. ea.,	240	Spin., 501 "	.838	6.704
16 " " 5 in. × 10 in., 40 "	640	" 601 "	.949	15.184
30 " " 4 in. × 8 in., 64 "	1,920	" 906 "	1.218	36.54
198 Throstle Spinning Fr., 128 "	25,344	" 3,700 "	1.5	297
6 Parr & Curtis Mules, 672 "	4,032	....	3	18
12 Filling Winders, 100 "	1,200	2,010 "	1.442	17.304
13 Spoolers, 80 "	1,040	600 "	.342	4.446
8 Reels,	.....	....	.143	1.144
16 Warpers,	.....	....	.171	2.736
3 Slashers,	.....	....	1.581	4.743
2 Dressers,	.....	....	1.141	2.282
2 Size Kettles,	.....	....	.153	.306
622 34-in. Looms,	.....	118 Picks per min.	.193	120.046
6 36-in. "	.....	118 " "	.193	1.158
60 38-in "	.....	118 " "	.197	11.82
217 40-in. "	.....	118 " "	.2	43.4
Total Machinery =	.....	....	.....	744.221
Shafting, New, at 10 %, =	.....	....	.....	74.442
Total =	.....	....	.....	818.643

H. P. per 1,000 Spindles = 27.85 = 35.9 Spindles per H. P.

Per cent. of Power of Machinery,
{ Picking and Carding, 29.54
Spinning, 42 33
Dressing, 4.42
Weaving, 23.71
——
100 }

SUMMARY OF POWER OF MILLS.—(Continued.)

MILL C.—*Fancy Pantaloonery, Shirting Stripes, etc. Average No. of Yarn, 16.5.*

Description	Total Spin.	Speed.	H.P. ea.	H. P.
1 Creighton Willow,	....	....	....	8
4 Kitson 2-Beater Openers,	....	Beat'rs, 1,500 Rev.	4.571	18.284
44 Breaker Cards,	....	Cylin'r, 120 "	.145	6.38
2 " Lap Heads, 22 Cards each,	....	10 yds. per min.	1.016	2.032
1 " " Doubler,	....	10 " "	....	.345
66 Finisher Cards,	....	Cylin'r, 120 Rev.	.145	9.57
6 " Railways,	....	F. Roll, 320 "	.38	2.28
6 Heads, 4 Roll 1st Drawing, 12 Deliv's,	....	" 155 "	del.121	1.454
6 " 4 " 2d " 12 "	....	" 200 "	".165	1.86
3 Slubbers, 64 Spin. each, 10 in. x 5 in.,	192	Spin., 634 "	1.682	5.046
5 Intermed'te,72 " 9 in. x 4 in.,	360	" 676 "	1.48	7.4
9 Fine F. Fr., 136 " 7 in. x 3 in.,	1,224	" 861 "	1.377	12.393
12 Ring Fr., 160 " No. 9 Yarn,	1,920	" 4,053 "	1.852	22.226
11 " 160 " " 16 "	1,760	5,067	1.927	21.197
10 " 160 " " 22 "	1,600	5,067	1.917	19.17
3 " 140 " " 22 "	420	5,067	1.666	5.068
8 Mas'n Mu.,576 " " 17 "	4,608	3,400	1.562	12.496
6 Reels, 60 in.,	....	160	.143	.858
6 Skein Spoolers, 60 in.,	....	....	.34	2.04
3 Bobbin Spoolers, 60 in.,	....	....	.25	.75
4 Filling Winders, 80 in.,	....	2,000	.878	3.312
7 Warpers,	....	....	.119	.833
4 Dressers,	....	8 yds. per min.	1.5	6
100 36-in. Plain Looms,	...	120 Picks	.158	15.8
36 40-in. "	....	120 Picks	.166	6
100 36-in. Crompton,	....	118	.234	23.45
Total Machinery =	....	....	....	214.244
Add Shafting, 10%,=	....	....	....	21.424
Total =	....	....	....	235.608

H. P. per 1,000 Spindles = 22.85 = 43.77 Spindles per H. P.

Per cent. of Power of Machinery,
- Picking and Carding, 35.03
- Spinning, 37.41
- Dressing, 6.44
- Weaving, 21.12
- 100

SUMMARY OF POWER OF MILLS.—(*Continued.*)

NEW MILL D.—*Fine Sheetings. Average Number of Yarn, 28.*

Description.	Total Spin.	Speed.	H.P. ea.	H. P.
1 Van Winkle Opener,	....	....	....	2
4 Platt's 36-in. 2-Beater Lappers,	....	Beat'rs, 1,100 Rev.	4.52	18.08
64 Saco 36-in. Breaker Cards,	....	Cylin'r, 125 "	.093	5.952
1 " " " Lap Head,	....	22 ft. per min.	....	1.02
64 " Finisher Cards,	....	Cylin'r, 125 Rev.	.093	5.952
8 " " Railways,	....	F. Roll, 230 "	.233	1.864
8 Heads, 5-Roll Drawing, 32 Deliv's,	....	" 196 "	del .083	2.648
2 Slubbers, 56 Sp. ea., 12 in. x 6 in.,	108	Spin., 530 "	1.259	2.518
4 Intermediates, 88 " 10 in. x 5 in.,	352	" 630 "	1.091	4.364
8 Fine F. Fra., 152 " 7 in. x 3 in.,	1,216	" 1,060 "	1.256	10.048
53 R'g Spin. Fra., 144 "	7,632	" 4,972 "	1.326	70.808
8 Mules, 560 "	4,480	" 3,812 "	1.736	13.888
4 " 592 "	2,368	" 3,812 "	1.835	7.34
5 Spoolers, 100 "	500	" 700 "	.327	1.635
5 Warpers,	....	....	.125	.625
1 Slasher,	....	....	....	1.5
180 36-in. Looms,	....	125 Picks	.108	19.44
136 40-in. "	....	125 "	.116	15.776
Total Machinery =	....	....	....	185.458
Shafting by tests =	....	....	....	21.5
Total =	....	....	....	206.958

H. P. per 1,000 Spindles = 14.25 = 70 Spindles per H. P.

Per cent. of Power of Machinery,
{ Picking and Carding, 29.35
  Spinning,   49.64
  Dressing,   2.01
  Weaving,   19
                       ———
                       100 }

MILL E.—*Fine Sheetings. Average Number of Yarn, 32.*

Description.	Total Spin.	Speed.	H.P. ea.	H. P.
1 Opener and Mixer,	....	Beatr's, 700 Rev.	....	2
2 2-Beater Whitin Pickers,	....	" 2,000 "	3	6
2 3-Beater " "	....	" 2,000 "	4	8
48 36-in. Cards,	....	Cylin'r, 130 "	.288	13.824
2 Railways to same,	....	11⅜ yds. per min.	.247	.494
1 Doubler "	....	....	....	.25
48 36-in. Finisher Cards,	....	Cylin'r, 132 Rev.	.207	9.868
4 Railways to same,	....	F. Roll, 394 "	.507	2.028
1 Frame, 1st Drawing, 5-Roll, 6 Deliv's,	....	" 381 "	.19	1.138
1 " 2d " " 8 "	....	" 361 "	.207	1.655

SUMMARY OF POWER OF MILLS.—(*Continued.*)

Mill E.—(*Continued.*)

Description.	Total Spin.	Speed.	H.P. ea.	H. P.
1 Frame, 3d Drawing, 5-Roll, 20 Deliv's,	....	F. Roll, 380 Rev.	.097	1.907
5 Slubbers (Higgins), Bob. 9 in. × 4.2 in.,	352	Spin., 530 "	.581	2.944
5 Fine Fr., " " 7 in. × 3.2 in.,	600	" 750 "	.982	4.91
6 " (Hill) " 7 in. × 3.2 in.,	680	" 750 "	.824	4.946
52 Ring Frames,	7,328	" 5,800 "	1.74	90.479
16 Marvel & Davol Mules,	9,024	" 4,350 "	2.051	32.814
4 Spoolers,	360	" 2,000 "	.327	1.31
4 Warpers,	....	....	.125	.5
1 Slasher,	....	....	....	1.061
343 40-in. Looms,	....	129 Picks per min.	.111	37
Total Machinery =	....	....	....	223.178
Shafting, 10% , =	....	....	....	22.318
Total H. P. =	....	....	....	245.496

H. P. per 1,000 Spindles = 15.01 = 66.6 Spindles per H. P.

Per cent. of Power of Machinery, { Picking and Carding, 26.88 / Spinning, 55.25 / Dressing, 1.29 / Weaving, 16.58 } 100

New Mill F.—*Fine Shirtings and Cambrics. Average Number of Yarn, 33.*

Description.	Total Spin.	Speed.	H.P. ea.	H. P.
1 Kitson Opener,	....	....	....	6
4 " 2-Beater Lappers,	....	Beat'rs, 1,500 Rev.	4.5	18
60 Breaker Cards,	....	Cylin'r, 120 "	.125	7.5
1 " Lap Head,	....	....	....	1
60 Finisher Cards,	....	" 120 "	.125	7.5
6 " Railways,	....	....	.5	3
2 Frames, 4-Roll Drawing, 12 Deliv's,	....	F. Roll, 221 "	del.096	1.16
2 " " " 12 "	....	" 221 "	".096	1.16
4 Slubbers, 60 Sp. ea., 10 in. × 5 in.,	240	Spin., 615 "	.789	3.156
6 Intermediates, 80 " 9 in. × 4½ in.,	480	" 773 "	1.238	7.428
11 Fine F. Fr., 144 " 7 in. × 3¼ in.,	1,584	" 934 "	1.05	11.55
64 Ring Sp. Fr., 128 "	8,192	" 5,908 "	1.258	80.612
14 F.F. Pat. Mu., 704 "	9,856	" 5,000 "	2.084	29.176
4 Spoolers, 100 "	400	" 600 "	.25	1
6 Warpers,	....	....	.125	.75
1 Slasher,	....	....	....	1.5
200 40-in. Looms, Heavy Cloth,	....	120 Picks	.2	40
200 40-in. " Light "	....	120 "	.135	27
Total Machinery =	....	....	....	247.492
Shafting, etc., 10% , =	....	....	....	24.749
Total Power =	....	....	....	272.241

SUMMARY OF POWER OF MILLS.—(Continued.)

MILL F.—(Continued.)

H. P. per 1,000 Spindles = 15.08 = 66.31 Spindles per H. P.

Per cent. of Power of Machinery,
{ Picking and Carding, 27.25
Spinning, 44.32
Dressing, 1.36
Weaving, 27.07
———
100 }

MILL G.—*Old, partially renewed, on Corset Jeans. Average No. of Yarn, 33.*

Description.	Total Spin.	Speed.		H.P. ea.	H. P.
1 Single Creighton Willow,	....	Beat'rs,	820 Rev.	....	5.402
2 Platt's 2-Beater, 48-in., 1st Pickers,	....	"	1,016 "	4.848	9.696
2 " " " 2d "	....	"	1,066 "	4.566	9.132
68 Old 24-in. Breaker Cards,	....	Cylin'r,	127 "	.185	12.92
1 Lap Head to same, 36 in.,	....	11 yds. per min.		....	2.539
44 36-in. Finisher Cards,	....	Cylin'r,	127 "	.268	11.792
4 Railways to same,	....	F. Roll,	360 "	.512	2.048
4 Heads, 5-Roll Drawing, 12 Deliv's,	....	"	258 "	del .144	1.732
4 " " " 16 "	....	"	296 "	".136	2.182
2 Slubbers, 60 Sp. ca., 12 in. × 6 in.,	120	Spin.,	543 "	1.448	2.896
6 Intermediat's, 80 " 9 in. × 4.2 in.,	480	"	630 "	.977	5.862
12 Fine F. Fr., 136 " 6 in. × 3 in.,	1,632	"	1,000 "	.983	11.796
36 R'g Sp. Fr., 192 "	7,056	"	6,000 "	2.143	77.148
12 Mason Mules, 512 "	6,144	"	3,690 "	2.135	25.62
8 " " 480 "	3,840	"	3,690 "	2.045	16.36
6 Spoolers, 80 "	480	"	600 "	.228	1.368
4 Warpers,	....	....		.177	.71
1 Slasher,	....	....		....	1.5
100 36-in. Looms,	....	130 Picks		.104	10.4
223 40-in. "	....	130 "		.135	30.077
28 48-in. "	....	122 "		.138	3.864
Total Machinery =	....	....		....	245.044
Shafting, etc., from tests, =	....	....		....	27.41
Total H. P. =	....	....		....	272.454

H. P. per 1,000 Spindles = 16 = 62.5 Spindles per H. P.

Per cent. of Power of Machinery,
{ Picking and Carding, 31.83
Spinning, 48.61
Dressing 1.46
Weaving, 18.1
———
100 }

SUMMARY OF POWER OF MILLS.—(*Continued.*)

MILL H.—*New, on Print Cloth, all Mule Spinning. Average No. of Yarn,* 31.

Description.	Total Spin.	Speed.	H.P. ea.	H. P.
1 Double Creighton Willow,	....	....	....	12
4 Platt Bros.' 1st Pickers,	....	Beat'rs, 1,130 Rev.	5.806	23.544
4 " 2d " with eveners,	....	" 1,130 "	6.256	25.024
176 36-in. Cards,	....	Cylin'r, 136 "	.167	29.44
12 Railways to same,	....	F. Roll, 412 "	.689	8.273
4 Frames, 4-Roll, 1st Draw'g, 32 Deliv's,	....	" 233 "	del.115	3.696
4 " " 2d " 48 "	....	" 238 "	".108	5.184
4 Slubbers, 56 Sp. ea.,	224	Spin., 630 "	1.567	6.268
2 " 48 "	96	" 630 "	1.318	2.636
14 Intermediates, 66 "	924	" 694 "	1.418	19.852
20 Fine F. Frames, 160 "	3,200	" 1,070 "	2.57	51.4
38 Warp Mules, Platt Bros., 552 "	20,976	" 5,300 "	3.594	136.572
34 Weft " " 600 "	20,008	" 4,200 "	2.732	92.888
14 Spoolers, 80 "	1,120	" 630 "	.186	2.6
12 Warpers,	....	....	.113	1.36
3 Slashers,	....	....	1	3
1,008 Looms,	....	154 Picks	.103	103.549
Total Machinery =	....	....	....	527.286
Shafting, etc., 10%, =	....	....	....	52.728
Total H. P. =	....	....	....	580.014

H. P. per 1,000 Spindles = 13.88 = 72 Spindles per H. P.

Per cent. of Power of Machinery,
{ Picking and Carding, 35.52
  Spinning, 43.52
  Dressing, 1.32
  Weaving, 19.54
  ——
  100 }

SUMMARY OF POWER OF MILLS.—(*Continued.*)

Mill I.—*New, on Print Cloths, all Mule Spinning. Average No. of Yarn*, 32.

Description.	Total Spin.	Speed.	H.P. ea.	H. P.
2 Kitson's Compound Openers,	....	Beat'rs, 1,500 Rev.	11	22
3 " 2-Beater Pickers,	....	" 1,500 "	5.536	16.608
144 36-in. Cards,	....	Cylin'r, 120 "	.227	32.688
12 Railway Heads,	....	....	.666	8
24 Deliveries, 1st Drawing,	....	F. Roll, 220 "	del .131	3.136
48 " 2d "	....	" 210 "	" .078	3.72
8 Slubbers, 48 Sp. ea., 12 in. × 6 in.,	384	Spin., 550 "	1.4	11.2
12 Intermediates, 68 " 10 in. × 5 in.,	816	" 650 "	1.707	20.484
20 Fine F. Fra., 144 " 7 in. × 3.2 in.,	2,880	" 1,060 "	1.808	36.16
32 Parr & Curtis Warp Mules, 564 Sp. ea.,	18,048	" 5,110 "	3.025	96.8
28 " Weft " 600 "	16,800	" 4,110 "	2.287	64.036
8 Spoolers, 130 "	1,040	" 640 "	.192	1.536
8 Warpers,	....	....	.366	2.928
2 Slashers,	....	....	.702	1.404
800 Looms,	....	154 Picks	.115	92.24
Total Machinery =	....	....	....	408.94
Shafting, etc., 10%, =	....	....	....	40.894
Total H. P. =	....	....	....	449.834

H. P. per 1,000 Spindles = 13.03 = 76.74 Spindles per H. P.

Per cent. of Power of Machinery,
{ Picking and Carding, 36.68
Spinning, 39.33
Dressing, 1.43
Weaving, 22.56
———
100 }

Indicator Cards of Engine, 470.57 H. P.
Less 5% for Engines, 23.52 "

Net H. P. of Mill =   447.05

SUMMARY OF POWER OF MILLS.—(*Continued.*)

Mill K.—*New, on Fine Cambrics. Average Number of Yarn,* 49.

Description.	Total Spin.	Speed.	HP. ea.	H. P.
1 Kitson 2-Beater Opener,	....	Beat'rs, 1,350 Rev.	....	6
4 " " Lappers,	....	" 1,350 "	3.776	15.104
52 36-in. Breaker Cards,	....	Cylin'r, 128 "	.085	4.42
1 Breaker Lap Head.	....	9 yds. per min.	....	1.437
52 36-in. Finisher Cards,	....	Cylin'r, 128 Rev.	.129	6.708
4 " Railways to same,	....	F. Roll, 290 "	.53	2.12
12 Deliveries, 1st Drawing, 5-Roll,	....	" 226 "	del.11	1.324
16 " 2d " "	....	" 226 "	.105	1.684
3 Slubbers, 48 Sp. ea., 12 in. × 6 in.,	144	Spin., 590 "	1.343	4.029
5 Intermediates, 80 " 10 in. × 5 in.,	400	" 736 "	1.3	6.5
5 Fine F. Fra., 136 " 7 in. × 3.5 in.,	680	" 968 "	1.496	7.48
4 " " 136 " 7 in. × 8.5 in.,	544	" 979 "	1.302	5.208
14 Jack Frames, 144 " 5 in. × 2 in.,	2,016	" 1,117 "	1.096	15.344
61 R'g Spin. Fr., 160 "	9,760	" 5,800 "	1.084	68.292
8 Parr & Curtis Mules, 696 Sp. ea.,	5,568	" 5,600 "	3.476	27.808
8 " " 552 "	4,416	" 5,600 "	3.222	25.776
4 Spoolers, 100 "	400	" 660 "	.167	.668
4 Warpers,	....	....	.118	.472
1 Slasher,	....	....	....	1.555
288 Looms (New),	....	150 Picks	.137	39.456
Total Machinery =	....	....	....	241.385
Shafting, etc., 10%, =	....	....	....	24.138
Total H. P. =	....	....	....	265.523

H. P. per 1,000 Spindles = 13.28 = 75.58 Spindles per H. P.

Per cent. of Power of Machinery,
{ Picking and Carding, 32.04
  Spinning, 50.5
  Dre·sing, 1.12
  Weaving, 16.34
  ——
  100

SHAFTING.

The writer is indebted to the courtesy of James B. Francis, Esq., of Lowell, for permission to copy the formulæ and tables prepared by him for the strength and velocity of shafting, after long and careful tests made for the Merrimac Manufacturing Co., and originally published by him in "The Journal of the Franklin Institute" for 1867, viz.:

For 1st shafts, or prime movers, subject to the strain of gears or main pulleys:

$$\text{Wrought Iron. Diam.} = \sqrt[3]{\frac{100 \times \text{H. P. to be transmitted}}{\text{No. of Rev. per min.}}}$$

(The breaking strain being taken at 56,000 lbs. per sq. in.)

$$\text{Cast Iron. Diam.} = \sqrt[3]{\frac{167 \times \text{H. P. to be transmitted}}{\text{No. of Rev. per min.}}}$$

(Breaking strain taken at 30,000 lbs. per sq. in.)

$$\text{Steel. Diam.} = \sqrt[3]{\frac{62.5 \times \text{H. P. to be transmitted}}{\text{No. of Rev. per min.}}}$$

(Breaking strain taken at 80,000 lbs. per sq. in.)

Being equal to $15\frac{1}{2}$ times the breaking power.

For 2d movers, or long lines, transmitting power:

$$\text{Wrought Iron. Diam.} = \sqrt[3]{\frac{50 \times \text{H. P.}}{\text{No. Rev. per min.}}}$$

$$\text{Cast \quad `` \quad `` } = \sqrt[3]{\frac{83 \times \text{H. P.}}{\text{No. Rev. per min.}}}$$

$$\text{Steel. \quad `` } = \sqrt[3]{\frac{31.25 \times \text{H. P.}}{\text{No. Rev. per min.}}}$$

Being equal to $7\frac{3}{4}$ times the breaking power.

For 3d movers, or light counters, driving machines well supported by bearings at short distances apart:

$$\text{Wrought Iron. Diam.} = \sqrt[3]{\frac{33 \times \text{H. P.}}{\text{No. Rev. per min.}}}$$

$$\text{Cast \quad `` \quad `` } = \sqrt[3]{\frac{55.5 \times \text{H. P.}}{\text{No. Rev. per min.}}}$$

$$\text{Steel. \quad `` } = \sqrt[3]{\frac{21 \times \text{H. P.}}{\text{No. Rev. per min.}}}$$

Or $5\frac{1}{4}$ times the breaking power.

From the above formulæ the relative diameters necessary for the same strength may be obtained as follows :

Wrought Iron 1
Cast   " 1.184
Steel 0.855

And the necessary size for shafts of the latter materials may be calculated from the following tables for wrought iron, which the writer has computed from Mr. Francis's data to an extent covering all necessary demands.

These tables apply to the torsional strains, but it is often necessary to use shafts larger than are required to transmit the power, in order to avoid the transverse strain and consequent friction due to flexure, in regard to which I quote Mr. Francis as follows :

Table of the greatest admissible Distances between the Bearings of continuous Shafts, subject to no Transverse Strains except from their own Weight.

DIAMETER OF SHAFT, IN INCHES.	DISTANCE BETWEEN BEARINGS, IN FEET.	
	If of Wrought Iron.	If of Steel.
1	12.27	12.61
2	15.46	15.89
3	17.7	18.19
4	19.48	20.02
5	20.99	21.57
6	22.3	22.92
7	23.48	24.13
8	24.55	25.23
9	25.53	26.24
10	26.44	27.18
11	27.3	28.05
12	28.1	28.88

" In practice long shafts are scarcely ever entirely free from transverse strains ; however, in the parts of long lines which have no pulleys or gears, with the couplings near the bearings, the interval between the bearings may approach the distances given in the preceding table. Near the extremities of a line the distances between the bearings should be less than are given in the table. The last space should not exceed sixty per cent. of the distance there given, the deflection in that space being much greater than in other parts of the line. In shafts moving with high velocities it will usually be necessary to shorten the distances between the bearings as given in the table, in order to obtain sufficient bearing surface to prevent heating.

"In factories and workshops power is usually taken off from the lines of shafting at many points by pulleys and belts, by means of which the machinery is operated. When the machines to be driven are below the shaft, there is a transverse strain on the shaft due to the weight of the pulley and tension of the belt, which is in addition to the transverse strain due to the weight of the shaft itself. Sometimes the power is taken off horizontally on one side, in which case the tension of the belt produces a horizontal transverse strain, and the weight of the pulley acts with the weight of the shaft to produce a vertical transverse strain. Frequently the machinery to be driven is placed above the floor, to which the shaft is hung in the story below; in this case the transverse strain produced by the tension of the belt is in the opposite direction to that produced by the weight of the pulley and shaft. Sometimes power is taken off in all these directions from the part of a shaft between two adjacent bearings. To transmit the same power the necessary tension of a belt diminishes in proportion to its velocity; consequently, with pulleys of the same diameter, the transverse strain will diminish in the same ratio as the velocity of the shaft increases. In cotton and woolen factories with wooden floors the bearings are usually hung on the beams, which are usually about eight feet apart; and a minimum size of shafting is adopted for the different classes of machinery, which has been determined by experience as the least that will withstand the transverse strain. This minimum is adopted independently of the size required to withstand the torsional strain due to the power transmitted; if this requires a larger diameter than the minimum, the larger diameter is of course adopted. In some of the large cotton factories in this neighborhood, in which the bearings are about 8 feet apart, a minimum diameter of $1\frac{7}{8}$ inch was formerly adopted for the lines of shafting driving looms. In some mills this is still retained; in others $2\frac{1}{8}$ inches and $2\frac{3}{16}$ inches have been substituted. In the same mills the minimum size of shafts driving spinning machinery is from $2\frac{1}{8}$ to $2\frac{11}{16}$ inches. In very long lines of small shafting fly-wheels are put on at intervals, to diminish the vibratory action due to the irregularities in the torsional strain."

The proper velocity for shafting has been of late the subject of much discussion and experiment, and has been greatly increased from former standards in the most approved modern mills, and a velocity of from 200 to 250 revolutions per minute is now usually adopted for carding and weaving rooms, as giving a fair proportionate size of pulleys, where the speed of the pulleys on the cards and looms varies from 130 to 160 revolutions per minute, while 300 to 350 revolutions per minute seems not too much for spinning rooms, where the speed of the cylinder on the frames varies from 600 to 900 revolutions, and will allow the use of 20-inch to 30-inch pulleys on the shaft, belting on to a 10-inch pulley on the frame, instead of the 6-inch or 7-inch pulleys formerly used, thus giving a much better holding surface to the belt, and from its high velocity allowing it to be much lighter while transmitting the same power, and with less strain and friction on the journals. At these velocities a line of 2-inch shaft in the weaving room, at 200 revolutions, will transmit 32 horse-power, or drive 256 looms at 8 looms per horse-power, and one of $2\frac{1}{4}$ inches, at 300 revolutions, in the spinning room, will transmit 68 horse-power, or drive 6,800 spindles at 100 spindles per horse-power.

These sizes are, however, capable of transmitting all the power required for the whole line, and are usually diminished as the power is taken off at intervals; but in such cases care is taken either to place the transmitting pulleys as close as possible to the bearings or to add supplementary bearings to support the shaft close to the pulleys, where it is necessary to place the latter in or near the middle of a "bay," or space between beams. In some of the latest mills the sizes given by the table for 3d movers are adopted for line shafting or 2d movers, and additional hangers for bearings are provided; and in one new mill which the writer has recently visited the main tie beams of the mill are 10 feet apart, but the bays or spaces are divided by supplementary beams for the support of bearings, so that the shafting is everywhere supported at intervals of 5 feet. This gives it a sufficient resistance to flexure to permit of the use of the third table of sizes, in which the factor of safety or strength in excess of the breaking strain is $5\frac{1}{6}$, which is ample for most purposes so far as strength is concerned, although machines having a reciprocating motion, like looms and mules, will sometimes require a greater diameter, to insure rigidity of shaft and steadiness of motion.

The use of "cold-rolled shafting" will also enable the further application of the third table to sizes for 2d movers or line shafting, as the experiments made by Professor Thurston, of the Stevens Institute of Technology, on cold-rolled iron from the works of Messrs. Jones & Loughlin, of Pittsburgh, Pa., show its great advantage in stiffness and elasticity, as might be expected from the perfect and uniform condensation of the fibres of the iron. We have not space here to copy the details of the experiments, but give the conclusions drawn by Professor Thurston, stating also that our own observation of mills where this shafting has been introduced leads us to believe in its great superiority to turned shafting from hot-rolled iron. The writer would not, however, advise the use of any line shafting less than $1\frac{1}{2}$ inch diameter, except possibly the last length in the line.

1. The process of cold rolling produces a very marked change in the physical properties of the iron thus treated:

(*a.*) It increases the tenacity from 25 to 40 per cent., and the resistance to transverse stress from 50 to 80 per cent.

(*b.*) It elevates the elastic limits under both tensile and transverse stresses from 80 to 125 per cent.

(*c.*) The modulus of elastic resilience is elevated from 300 to 400 per cent. The elastic resilience to transverse stress is augmented from 150 to 425 per cent.

2. Cold rolling also improves the metal in other respects :

(*a.*) It gives the iron a smooth, bright surface, absolutely free from the scale of black oxide unavoidably left when hot rolled.

(*b.*) It is made exactly to gauge, and for many purposes requires no further preparation.

(*c.*) In working the metal the wear and tear of the tools are less than with hot-rolled iron, thus saving labor and expense in fitting.

(*d.*) The cold-rolled iron resists stresses much more uniformly than does the untreated metal. Irregularities of resistance exhibited by the latter do not appear in the former ; this is more particularly true for transverse stress, as is shown by the smoothness of the strain-diagrams produced by the cold-rolled bars.

(*e.*) This treatment of iron produces a very important improvement in uniformity of structure, the cold-rolled iron excelling common iron in its uniformity of density from surface to center, as well as in its uniformity of strength from outside to the middle of the bar.

The proportion of length of the bearing of a shaft to its diameter is a question which has caused much discussion, and the writer has been asked to give his opinion ; and, although he does it with all modesty, he will say that he is inclined to favor a length of three times the diameter, as being the best point for practical use when the shafts can be kept well in line and well lubricated. This proportion with proper couplings will afford sufficient bearings, and by the use of swivel bearings avoid any unnecessary twist or strain at the end of the boxes. Lubrication also deserves some notice, and here the author's opinions are positive, and confirmed by his tests, in favor of continuous lubrication with oil, which oil should be mixed to suit the weight in the bearing, in various proportions of mineral and animal oils. One half of each is a very good proportion for medium shafting, say petroleum and sperm or lard, while for light bearings the petroleum may be three fourths, and for heavy ones the animal oil may be neatsfoot.

Grease or tallow is an abomination ; and where old boxes fitted for it are in use, with holes through the cover of bearing for the tallow to run down when it gets hot enough to melt, these holes may be filled with sponge and kept saturated with oil.

Table of Horse-Power which can be safely carried by 1st Movers at different Velocities, Factor of Safety being = 15.5.

Diameter in Inches.	Revolutions per Minute.					Diameter in Inches.	Revolutions per Minute.			
	50	100	150	200	250	300		50	100	150	200
	Horse-Power.						Horse-Power.			
1	.5	1	1.5	2	2.5	3	6	108	216	324	432
1.25	.975	1.95	2.92	3.9	4.87	5.85	6.25	122.07	244.14	366.21	488
1.5	1.68	3.37	5.04	6.74	8.4	10.11	6.5	137.31	274.62	411.93	549
1.75	2.68	5.36	8.04	10.72	13.4	16 08	6.75	153.72	307.55	461.16	615
2	4	8	12	16	20	24	7	171.5	343	514.5	686
2.25	5.69	11.39	17.07	22.78	28.45	34.17	7.25	190.54	381.08	571.62	762
2.5	7.81	15.62	23.43	31.24	39.05	46.86	7.5	210.93	421.87	632.79	843
2.75	10.4	20.8	31.2	41.6	52	62.4	7.75	232.74	465.48	698.22	931
3	13.5	27	40.5	54	67.5	81	8	256	512	768	1,024
3.25	17.16	34.33	51.5	68.66	85.8	103	8.25	280.76	561.52	842	1,123
3.5	21.43	42.87	64.29	85.74	107.15	128.61	8.5	307.06	614.12	921	1,228
3.75	26.36	52.73	79.08	105.46	131.8	158.19	8.75	334.96	669.92	1,005	1,340
4	32	64	96	128	160	192	9	364 5	729	1,093.5	1,458
4.25	38.38	76.77	115.15	153.54	191.9	230.31	9.25	395.72	791.45	1,187	1,583
4.5	45.56	91.12	136.68	182.24	227.8	273.36	9.5	428.68	857.37	1,286	1,715
4.75	53.58	107.17	160.75	214.34	267.9	321.54	9.75	463.43	926.86	1,390	1,854
5	62.5	125	187.5	250	312.5	375	10	500	1,000	1,500	2,000
5.25	72.35	144.7	217.05	289.4	361.75	434.1	11	665.5	1,331	1,995	2,662
5.5	83.13	166.37	249.54	322.75	415.9	499.11	12	864	1,728	2,592	3,456
5.75	95.05	190.11	285.15	380.22	475.25	570.83	13	1,053.5	2,107	3,160.5	4,214

Table of Horse-Power for Shafting for Long Lines of Transmission, or 2d Movers, Factor of Safety being = 7.75.

Diameter in Inches.	Revolutions per Minute.				Diameter in Inches.	Revolutions per Minute.			
	100	150	200	250	300		100	150	200	250	300
	Horse-Power.					Horse-Power.				
1	2	3	4	5	6	3⅜	77.88	116.83	155.76	194.7	233.64
1⅛	2.88	4.32	5.76	7.2	8.64	3½	85.74	128.61	171.48	214.35	257.22
1¼	3.9	5.85	7.8	9.75	11.7	3⅝	95.25	142.86	190.5	238.1	285.75
1⅜	5.2	7.8	10.4	13	15.6	3¾	105.46	158.19	210.92	263.65	316.38
1½	6.74	10.11	13.48	16.85	20.22	3⅞	116.37	174.54	232.74	290.9	349
1⅝	8.58	12.87	17.16	21.45	25.74	4	128	192	256	320	384
1¾	10.72	16.08	21.44	26.8	32.16	4⅛	140.38	210.57	280.76	351	421.14
1⅞	13.18	19.77	26.36	32.95	39.54	4¼	153.54	230.31	307	383.85	460.62
2	16	24	32	40	48	4⅜	167.48	251.22	335	418.7	502.48
2⅛	19.19	28.77	38.38	47.95	57.57	4½	182.24	273.36	364.48	455.6	546.72
2¼	22.78	34.17	45.56	56.95	68.34	4⅝	197.86	296.79	395.72	494.65	593.58
2⅜	26.79	40.17	53.58	66.95	80.37	4¾	214.34	321.54	428.68	535.85	643
2½	31.24	46.86	62.48	78.1	93.72	4⅞	231.71	347.55	463.42	579.25	695.73
2⅝	36.18	54.27	72.36	90.45	108.54	5	250	375	500	625	750
2¾	41.6	62.4	83.2	104	124.8	5¼	289.4	434.1	578.8	723.5	868.2
2⅞	47.52	71.28	95	118.8	141.56	5½	332.75	499	665.5	831.87	998.25
3	54	81	108	135	162	5¾	380.22	570.33	760.44	950.55	1,140.66
3⅛	61.02	91.53	122	152.55	183	6	432	648	864	1,080	1,296
3¼	68.66	103	137.32	171.65	206						

7

Table of Horse Power for Shafting for Counter Shafts, well supported, or 3d Movers. Factor of Safety = 5.17.

Diameter in Inches.	Revolutions per Minute.						
	100	150	200	250	300	350	400
	Horse-Power.						
1	3	4.5	6	7.5	9	10.5	12
1 1/16	3.59	5.37	7.18	9.95	10.77	12.53	14.36
1 1/8	4.27	6.54	8.54	10.9	12.81	15.26	17.08
1 3/16	5.02	7.53	10.04	12.55	15.06	17.57	20.08
1 1/4	5.85	8.77	11.7	14.62	17.55	20.47	23.4
1 5/16	6.78	10.17	13.56	16.95	20.24	23.73	27.12
1 3/8	7.79	11.67	15.58	19.45	23.37	27.23	31.16
1 7/16	8.91	13.35	17.82	22.25	26.73	31.15	35.64
1 1/2	10.11	15.16	20.22	25.27	30.33	35.38	40.44
1 9/16	11.44	17.16	22.88	28.6	31.32	40	45.76
1 5/8	12.87	19.29	25.74	32.15	38.61	45	51.48
1 11/16	14.41	21.6	28.82	36	43.23	50.4	57.64
1 3/4	16.08	24.12	32.16	40.2	48.24	56.28	64.32
1 13/16	17.86	26.79	35.72	44.65	53.58	62.51	71.44
1 7/8	19.77	29.64	39.54	49.4	59.31	69.16	79.08
1 15/16	21.81	32.7	43.62	54.5	65.43	76.3	87.24
2	24	36	48	60	72	84	96
2 1/16	26.32	39.48	52.64	65.8	79	92.12	105.28
2 1/8	28.78	43.17	57.56	71.95	86.34	100.73	115.12
2 3/16	31.4	47.1	62.8	78.5	94.2	109.9	125.6
2 1/4	34.17	51.25	68.34	85.42	102.51	119.6	136.68
2 5/16	37.09	55.63	74.18	92.72	111.27	129.81	148.36
2 3/8	40.18	60.27	80.36	100.45	120.54	140.63	160.72
2 7/16	43.44	65.16	86.88	108.6	130.32	152.04	173.76
2 1/2	46.87	70.8	93.74	117.17	140.61	164.04	187.48
2 9/16	50.46	75.69	100.92	126.15	151.38	176.61	201.84
2 5/8	54.27	81.4	108.54	135.67	162.81	189.54	217.08
2 11/16	58.23	87.35	116.46	145.57	174.69	203.8	232.92
2 3/4	62.4	93.6	124.8	156	187.2	218.4	249.6
2 13/16	66.74	100.11	133.48	166.85	200.22	233.59	266.96
2 7/8	71.28	106.92	141.56	176.95	213.84	249.49	285.12
2 15/16	76.04	114.96	152.08	190.1	228.12	266.14	304.16
3	81	120.5	162	202.5	243	283.5	324

The above tables are carried out to an extent beyond all probable need, but may possibly be useful in extreme cases; and it should be remembered that the first length of shaft in a line, which carries the receiving pulley, and has to bear the vertical or lateral strain of the main belt, being also usually of considerable length, should generally be of the size given in the first table.

BELTING.

Any general rule for the speed of belts to convey a given number of horse-powers will of course be somewhat varied by situation and circumstances, but the writer believes that the following data and deductions will be found reliable for well tanned leather belts under ordinary conditions:

Morin gives .551 lb. per .00155 sq. in. section as a safe working strain, which is equal to 551 lbs. per 1.55 sq. in., or 355 lbs. per sq. in., and assumes the thickness of an ordinary single belt to be .16 in., which gives the safe strain on each inch of width to be equal to 56.8 lbs.

Haswell, in his "Engineer's Pocket-Book," gives the safe strain in like manner at 350 lbs. per sq. in., or equal to 56 lbs. per inch width of ordinary belt.

Rankine gives 285 lbs. per sq. in., or 45.6 per inch width, and copies from Towne's tables, in "The Journal of the Franklin Institute," the following:

```
Breaking strain per inch width in solid leather.................. 675 lbs.
    "      "      "   "  at rivet holes of splice................. 362  "
    "      "      "   "  at lacing holes.......................... 210  "
Safe working tension .......................................... 45  "
```

Mr. James S. Atwood, of Wauregan, Conn., has prepared a table for his own use, based on 330 lbs. per sq. in. as a safe working tension.

The very valuable collection of data and observations published by Mr. J. H. Cooper, of Philadelphia, gives a very wide range of opinions from various authorities, extending from 40 to 100 lbs. per inch in width of ordinary belting, as consistent with safety.

Mr. Cooper has also published in "The Journal of the Franklin Institute" for November, 1878, a paper containing a translation from the French of M. Laborde, originally published prior to 1833, and based on a working tension of only 20 lbs. per inch in width, but from which Mr. Cooper deduces the following simple rule for strength, viz.: "*It is the stress in pounds which each inch of belt width will safely and continuously bear at any velocity.*"

The tests made with Riehle's breaking machine at the Centennial Exhibition showed a breaking strain per sq. in. ranging from 3,000 to 5,000 lbs., or from 500 to 833 lbs. per 1 in. in width and $\frac{1}{6}$ in. in thickness, which I assume to be about the average of single belting.

The writer's own experience has shown him that a rule given him many years since by an experienced mechanic, of "*600 ft. velocity per 1 inch of belt width per horse-power,*" was perfectly reliable; and the reasons for it may be deduced from the above data as follows:

Assuming as a basis a fair average from the various tests, of a safe working strain of 330 lbs. per sq. in., or 55 lbs. for $\frac{1}{6}$ in. in thickness, about one quarter of the strength shown by Mr. Towne's tests at the lacing holes, we may obtain a very simple formula for velocity: 33,000 lbs. lifted 1 ft. per minute being the accepted unit of a horse-power, 1 sq. in. of belt must then move 100 ft. per minute to transmit the same, 330×100 being $= 33,000$; and $\frac{1}{6}$ sq. in. or 1 in. width of ordinary belting must move 600 ft. per minute, equal to 50 sq. ft. of belt per minute, which I therefore adopt as my rule for single belts. Double belting will vary from $\frac{1}{4}$ to $\frac{1}{3}$ or $\frac{3}{8}$ in. in thickness, and of course require proportionately less velocity per horse-power; and the following rules may be deduced for all dimensions, viz.:

"Multiply the denominator of the fraction expressing the thickness of the belt in inches by 100, and divide by the numerator, for the necessary velocity in feet per minute for each inch in width;" viz., to transmit 1 horse-power:

$\frac{1}{6}$ in. $= 6 \times 100 = 600$ ft. per minute.

$\frac{3}{8}$ in. $= \dfrac{8 \times 100}{3} = 266.66$ "   "

$\frac{1}{4}$ in. $= 4 \times 100 = 400$ "   "

The velocity and width being given, to get the horse-power: "Divide the actual velocity by the velocity per horse-power as above, and multiply by the width;" viz., for a 12-in. belt, single, 2,400 ft. per minute:

$$\frac{2,400}{600} = 4 \times 12 = 48 \text{ H. P.}$$

The velocity and horse-power being given, to get the inches in width: "Divide the velocity by the velocity per inch obtained as above, and divide the horse-power by the product;" viz., for a belt 3,000 ft. per minute to transmit 50 horse-power :

$$\frac{3,000}{600} = 5. \quad \frac{50 \text{ H. P.}}{5} = 10 \text{ in.}$$

These rules will, however, be varied by circumstances. Belts, when stopping and starting, or shifting from one pulley to another, as in the case of looms and mules are frequent, should, on account of the wear and tear, be made wider than the power only requires. Also any great difference in the size of pulleys, materially decreasing the angle of friction on the smaller pulley, will require an increase of width, to give the necessary holding surface.

Belts should be used with the grain or hair side next the pulley ; they will hold better and wear longer.

So far as the capacity of the belt itself to transmit power is concerned, independent of the frictional surface of the pulleys, the following table may prove convenient for reference for single belts of the average thickness of $\frac{1}{4}$ in., and from it may be readily deduced the available power to be derived from double belts according to their thickness. There are, however, other points to be considered than the one of the actual strength of the belt, the most important one of which is its friction or "hold" upon the pulley.

It is generally conceded that the friction of a belt passing half around a pulley is equal to one half the strain on the belt ; or that an inch belt at 600 ft. per minute, with a strain of 55 lbs., would give a pressure of 27.5 lbs., and require a pulley which would give 1,200 lineal feet per minute of surface contact, to obtain the 1 H. P. to which the belt would be equal. Morin, in his "Mechanics," gives as the result of actual trials with a loaded belt over a wooden drum an average friction of 50 per cent., which would be increased by using a pulley covered with leather ; and a polished iron pulley, with a smooth, flexible belt, may, I think, be depended on in actual use for 50 per cent. The most scientific writers commit gross errors in treating of this question. Professor Rankine says that the rough or flesh side of the belt should be next to the pulley to get friction ; whereas the friction of a belt is due to close contact and the consequent atmospheric pressure from outside, so that the best result is obtained by a smooth surface of leather, which, being moderately elastic, admits of the complete expulsion of the air between the surfaces and the consequent full effect of the external pressure.

Considering this established in practice, that the *available* friction is 50 per cent. of the strain, I find in use the following rules (which agree very closely with my previous conclusions) for getting at the proper width of a belt where the speed and amount of contact surface are determined by the necessities of the case.

The Page Belting Co., of Concord, N. H., gives the following formula:

$$\text{Inches width} = \frac{\text{No. H. P.} \times 36{,}000}{\text{Velocity in ft.} \times \frac{1}{2} \text{ contact length in inches}}$$

In Cooper's admirable collection of "Belting Facts and Figures," I find the following: "Professor Thurston gives:

$$\text{Width in inches} = \frac{\text{No. H. P.} \times 7{,}000}{\text{Velocity in ft.} \times \text{contact length in ft.}}.$$

Mr. F. W. Bacon, C. E., says:

$$\text{Width in inches} = \frac{\text{No. H. P.} \times 6{,}000}{\text{Velocity in ft.} \times \text{contact length in ft.}}$$

which is only a different way of expressing the rule given by the Page Belting Co. Messrs. Hoyt Bros., of New York, say:

$$W = \frac{\text{H. P.} \times 5{,}334}{\text{Velocity} \times \text{contact in ft.}}.$$

Van Riper, of Paterson, gives the same rule; and some one, whose name is not given, says:

$$W = \frac{\text{H. P.} \times 26{,}000}{\text{Velocity} \times \text{contact in ft.} \times 6}."$$

My own deductions would give, in the terms of the Page formula:

$$\text{Width in inches} = \frac{\text{No. H. P.} \times 33{,}000}{\text{Velocity in ft.} \times \frac{1}{2} \text{ contact in inches}};$$

or, reduced to feet, as by Mr. Bacon, for a single belt:

$$W = \frac{\text{H. P.} \times 5{,}500}{\text{Velocity} \times \text{contact in ft.}};$$

or for a double belt:

$$W = \frac{\text{H. P.} \times 3{,}660}{\text{Velocity} \times \text{contact in ft.}}.$$

In this rule for double belts I have assumed $\frac{1}{4}$ in. thickness and $82\frac{1}{2}$ lbs. strain; but, if the belt be, as many are, $\frac{3}{8}$ in. thick, it would of course bear from 110 to 120 lbs., and 267 ft. per minute would give

1 H. P. per inch ; and the formula for contact with one half the surface, or 180°, would be

$$W = \frac{H. P. \times 2{,}444}{\text{Velocity} \times \text{contact length in ft.}}$$

These formulæ are based on my previous data of a velocity of 50 sq. ft. per minute, or a strain of 55 lbs. per inch on a single belt, and on the belt being in contact with one half the circumference of the pulley. Now, the friction varies with the arc of the circle with which the belt is in contact, and is only half as great on one quarter of a pulley as on one half of one ; so that double the surface in square inches will be required to transmit the same power in the former case that would be needed in the latter, and the numerator of the formula for single belts would be H. P. ×11,000. This will be easily understood by those who know the enormous hold given by passing the rope from a pulley-block once around a post where the whole surface is in contact. If one third of the circumference is in contact, the coefficient in the numerator would be 8,250.

Carrying out these rules, it will be easily seen that, where high speed is to be obtained by the use of small pulleys, a much greater width of belt is necessary to get the frictional surface than is called for by the strength of the leather ; and it will be found that for circular saws, cotton pickers, spinning frames, etc., a wider belt is needed than is due to the actual power transmitted. Take, for instance, a spinning frame, with a 7-in. pulley, 900 revolutions per minute, or 1,650 ft. belt velocity, and requiring 1½ H. P. One inch of belt at that speed would transmit 2¼ H. P., but the contact surface of the pulley would not be over 10 in. in length, and by the above rules calls for a 3-in. belt, which is the standard size for that purpose.

A good practical example of a main belt in actual use, under the writer's frequent observation, is that of a 24-in. double belt, at a velocity of 3,200 ft. per minute, transmitting 160 H. P. to a pulley 4 ft. 10 in. in diameter.

Taking the first formula for double belts as above, the width should be

$$W = \frac{160 \text{ H. P.} \times 3{,}660}{3{,}200 \times 7.58 \text{ (} \tfrac{1}{4} \text{ circle)}} = 24.14 \text{ in.}$$

This belt has now run seven years without repair.

According to the rules for the strength only, it would transmit 192 H. P., but the smaller pulley should then be 5 ft. 9.6 in. diameter, instead of 4 ft. 10 in.

Table of Power which may be transmitted by Single Belts of different Widths and Velocities, averaging one sixth of an inch in thickness.

Width of Belt.	Velocity in Feet per Minute.									
	600	800	1,000	1,200	1,500	2,000	2,500	3,000	3,500	4,000	5,000
	Horse-Power.									
1	1	1.33	1.66	2	2.5	3.33	4.16	5	5.83	6.66	8.33
2	2	2.66	3.33	4	5	6.66	8.33	10	11.66	13.33	16.66
3	3	4	5	6	7.5	10	12.5	15	17.5	20	25
4	4	5.33	6.66	8	10	13.33	16.66	20	23.33	26.66	33.33
5	5	6.66	8.33	10	12.5	16.66	20.83	25	29.16	33.33	41.66
6	6	8	10	12	15	20	25	30	35	40	50
8	8	10.66	13.33	16	20	26.66	33.33	40	46.66	53.33	66.66
10	10	13.33	16.66	20	25	33.33	41.66	50	58.33	66.66	83.33
12	12	16	20	24	30	40	50	60	70	80	100
14	14	18.66	23.33	28	35	46.66	58.33	70	81.66	93.33	116.66
16	16	21.33	26.66	32	40	53.33	66.66	80	93.33	106.66	133.33
18	18	24	30	36	45	60	75	90	105	120	150
20	20	26.66	33.33	40	50	66.66	83.33	100	116.66	133.33	166.66
22	22	29.33	36.66	44	55	73.33	91.66	110	128.33	146.66	183.33
24	24	32	40	48	60	80	100	120	140	160	200
26	26	34.66	43.33	52	65	86.66	108.33	130	151.66	173.33	216.66
28	28	37.33	46.66	56	70	93.33	116.66	140	163.33	186.66	233.33
30	30	40	50	60	75	100	125	150	175	200	250
32	32	42.66	53.33	64	80	106.66	133.33	160	186.66	213.33	266.66
34	34	45.33	56.66	68	85	113.33	141.66	170	198.33	216.66	283.33
36	36	48	60	72	90	120	150	180	210	240	300

WATER-WHEELS.

THE following tables of the Turbine Wheel tests at the Centennial Exhibition in Philadelphia in 1876 will be explained by the annexed extract from the Official Report made by me to Captain John S. Albert, Chief of the Bureau of Machinery. The calculations as published in the report were revised by another person after leaving my hands, and published over my name without my knowledge. Some errors were corrected, and many more introduced, and I have therefore recalculated the whole, with the aid of Vega's logarithmic tables ; so that the results as now shown may be regarded as substantially correct.

In several instances they have been confirmed by tests since made by other persons, to whom the wheels had been taken by disappointed exhibitors for new trials.

TESTS OF TURBINE WATER-WHEELS.

"The water was furnished by a pair of powerful centrifugal pumps, exhibited by Messrs. W. L. Andrews & Co., of New York, and driven by oscillating engines, which raised from 1,800 to 1,900 cubic feet of water per minute to a tank placed at the end of the Hydraulic Annex, the overflow of which was 33 feet above the level of the water in the large tank in the center of the building, from which it was pumped.

"This water usually formed the 'cataract,' which was stopped partially or wholly while testing the Turbines.

"From this tank a wrought-iron tube or 'penstock,' 4 feet in diameter, descended to the 'flume,' or case in which the wheels were set, and which was 8 feet in diameter by 6 feet in height, supported by a brick wall resting on a granite bedstone. From the wheels the water was conducted by an ample passage to a rack or strainer 30 feet from the wheel, and stretching across a brick tail-race 14 feet wide by 8 deep, at the lower end of which, 15 feet below the rack, was the measuring weir, 9 feet long, formed of a heavy cast-iron plate planed to a true edge one eighth of an inch thick, and beveled from that on the lower side at an angle of 45°. The upright ends of the weir were made of Georgia pine, cut and beveled to the same dimensions, and were carefully adjusted by Mr. Samuel S. Webber, and verified by myself.

"The hook gauge, loaned for the experiments by Mr. T. H. Risdon, was placed in a tight wooden box 6 feet up stream from the weir, and the water was admitted to this box, for the purpose of measurement of height, by a few $\frac{3}{8}$-inch holes bored in the bottom of the box, 3 feet below the surface of the water; and an examination of the very thorough test of the Tait wheel shows the sensitiveness with which the weir measurement responded to the changes of load and variation in the number of revolutions of the wheel.

"The apparatus for measuring the power consisted of a friction pulley fitted to the wheel shaft, 37.44 inches diameter and 18 inches face, which was clasped by a Prony brake, consisting of a pair of cast-iron shoes lined with wood, from one of which projected an oak arm 6 by 4 inches, through which a knife-edged eye-bolt was fastened at a distance from the center of the shaft of 10.5 feet, or the radius of a 66-feet circle. These portions of the apparatus, with the scale-pan and hydraulic regulator, 16 inches diameter, were also kindly loaned by Mr. Risdon.

"To facilitate the handling of the weights, this lever was connected by an iron rod with the short arm of a bell-crank or scale-beam 2 feet in height, while the longer arms, which were attached to the scale-pan and regulator, were 4 feet each, thus giving a leverage of 132 to 1 for each pound placed in the scale. All the pivots or bearings of this scale-beam were of steel, knife-edged, and bearing in hardened iron sockets.

"The weights used were United States standard, and were kindly loaned by Messrs. Fairbanks & Co. The pulley, weighing 1,000 pounds, rested on the shaft and step of the wheel, corresponding in some measure to the usual 'crown-gear'; but the brake, which weighed 1,600 pounds, was suspended by a swivel from a beam directly over the center of the wheels, so as to allow perfect freedom of motion in any direction. An examination of the records will also show the sensitiveness and accuracy of this part of the apparatus, every distance and dimension of which I carefully measured and adjusted personally before commencing the tests.

"The head of water acting on the wheels was ascertained by a gauge-rod, having a hook at the lower end, which was carefully kept at the level of the tail-water in a box sunk in the floor and connected with the tail-race by a perforated pipe; while a pipe led from the case to the level of the head-water, where a glass tube enabled the observer to read at once the acting head by the graduations on the upper end of the gauge-rod.

"Experiments not strictly belonging to the wheel tests were made, showing that the same wheel, with the same load, at different times repeated the number of revolutions very accurately, and proved the correctness of the apparatus. The revolutions of the wheel were ascertained by a worm-gear clock, which was thrown in and out of connection with the shaft of the wheel, at signals given by a bell, which was struck at intervals of one or two minutes, according to the length of test desired.

"The friction pulley was accurately balanced before commencing the tests, and, when the wheels themselves were truly set, ran with perfect steadiness and regularity.

"In conducting these tests I have been assisted by the following gentlemen, our watches being all set to the same time before commencing the tests, and simultaneous observations being taken during their entire duration. These observations being noted down as taken, a comparison of the different note-books gave a record of all the points in the test at every half-minute of its duration."

"Mr. Percy Sanguinetti read the hook gauge, giving the height of water on the weir; Mr. Philip R. Voorhees read the gauge giving the head of water acting on the wheel; Mr. Samuel S. Webber managed the counting clock and read the revolutions of the wheel, and also saw that the lubrication was perfect; while Mr. John Cotter, Superintendent of the Hydraulic Annex, kept the records of the weight and revolutions, and assisted me generally in various ways. I personally kept an eye on all points, and gave the bell-signals by which the observations were taken."

"Each exhibitor was allowed free access and liberty of observations during the tests of his own wheel; and, whatever may be the accuracy of the net results obtained, the comparative ones may be depended on, as the tests were all made under similar circumstances, and the different points watched and the notes taken throughout by the same observers, none of them having any interest whatever in the result, or any opportunity at the time of knowing what the observations were at other stations than their own."

"It is worthy of notice that the best results have been attained by wheels taken just as they came from the shop, without any especial finish or preparation, and the thoroughly exhaustive test of the Tait wheel is worth studying, as showing the accurate working of the apparatus."

"The Geyelin wheel, entered by R. D. Wood & Co., was so tightly fitted in the shop that I do not think we got a fair record of its power; and the Cope wheel used so much water that we could not carry the test out in full, but the percentage was gaining regularly up to the last trial, when we exhausted the supply of water, having reached over 1,860 cubic feet, or 14,000 gallons per minute.

"The Hunt wheel also taxed the supply of water to the utmost, and the third wheel from the York Company was only tested to prove or disprove what was believed to be an unsound principle, viz., that of shallow buckets and central discharge; and the result is confirmed by those obtained from some of the other wheels.

"The leakage of the flume was large during the first six trials, but by calking and tamping with lead was very much reduced at the test of the Tyler wheel, after which test the allowance was uniform of 14.352 cubic feet per minute waste to each wheel. In the first six tests it was taken as noted in the tables, and the amount is in all cases deducted from the water consumed per minute."

The temperature of the water until November 1 was 75° Fahr., giving a weight of 62.234 pounds per cubic foot. After that date it was taken at 70°, or 62.3 pounds per cubic foot.

TESTS OF WATER-WHEELS—INTERNATIONAL EXHIBITION, 1876.

September 18. Barker & Harris, Turbine. 20 Inches Diameter.

No. of Test.	Time of Start.	Time of Stop.	Weight Lift'd.	Revolutions per Min.	Head on Wheels.	Head on Weir	Cubic Feet Discharged per Min.	Horse-Power of Wheel.	Horse Power of Water.	Percentage of Effect.	Waste on Weir.	REMARKS.
	P.M.	P.M.	Lbs.									
1	5.05	5.07	23	354	31.25	.02	812.43	32.57	47.88	.6302	.096	Full Gate.
2	5.08	5.10	26	348.5	31.22	.623	818.66	36.244	48.2	.7519	.096	"
3	5.13	5.15	27	341.5	31.18	.63	833.28	36.882	49	.7527	.096	"
4	5.18	5.19	28	330.5	31.18	.626	824.88	37.02	48.5	.7631	.096	"
5	5.21	5.23	22	380.5	31.27	.6	771.22	33.484	45.48	.7362	.096	¾ Gate.
6	5.27	5.29	22	287.5	31.4	.514	601.72	25.3	35.63	.71	.096	½ "
7	5.30	5.32	20	299	31.45	.495	566	23.92	33.57	.7125	.096	½ "
8	5.40	5.42	16	271.5	31.62	.405	405.97	17.376	24.21	.7177	.096	¼ "
9	5.47	5.49	13	327.5	31.66	.405	405.97	17.03	24.24	.7026	.096	¼ "

Waste, .096 = 53.34 cu. ft. per minute, deducted from cu. ft. per minute gross, to give amount in table.

September 21. Risdon Wheel. 30 Inches Diameter.

No. of Test.	Time of Start.	Time of Stop.	Weight Lift'd.	Revolutions per Min.	Head on Wheels.	Head on Weir.	Cubic Feet Discharged per Min.	Horse-Power of Wheel.	Horse-Power of Water.	Percentage of Effect.	Waste on Weir.	Remarks
	P.M.	P.M.	Lbs.									
1	1.07	1.09	78	266	30.36	.973	1,653.85	82.99	94.69	.8768	.072	Full Gate.
2	1.10	1.12	80	258.5	30.36	.9795	1,669.31	82.73	95.57	.8655	.072	"
3	1.13	1.15	82	252.5	30.37	.9804	1,671.27	82.82	95.72	.8652	.072	"
4	1.18	1.20	68	257	30.59	.8738	1,403.67	69.9	80.96	.8622	.072	⅞ Gate.
5	1.21	1.23	70	247	30.59	.876	1,410.94	69.16	81.36	.85	.072	⅞ "
6	1.26	1.28	60	238	30.83	.795	1,210.38	57.12	70.41	.8112	.072	¾ "
7	1.31	1.33	58	248	30.84	.7876	1,198.92	57.53	69.81	.8241	.072	¾ "
8	1.38	1.40	38	269	31.05	.677	951.81	40.88	55.74	.7316	.072	½ "
9	1.41	1.43	40	263.5	31.04	.68	958.44	42.16	56.11	.7513	.072	½ "
10	1.44	1.46	41	258	31	.681	960.54	42.31	56.15	.7535	.072	½ "

Waste on weir, .072 = 34.66 cu. ft. per minute, deducted from gross amount, to give amount in table.

September 23. Knowlton & Dolan. 24 Inches Diameter.

No. of Test.	Time of Start.	Time of Stop.	Weight Lift'd.	Revolutions per Min.	Head on Wheels.	Head on Weir.	Cubic Feet Discharged per Min.	Horse-Power of Wheel.	Horse-Power of Water.	Percentage of Effect.	Waste on Weir.	Remarks.
	P.M.	P.M.	Lbs.									
1	12.26	12.28	50	332.5	30.82	.908	1,482.3	66.7	86.13	.7743	.082	Full Gate.
2	12.29	12.31	52	324	30.79	.9195	1,510.89	67.39	87.73	.7681	.082	"
3	12.38	12.40	54	311	30.75	.923	1,519.7	67.17	88.13	.7622	.082	"
4	12.41	12.43	56	302	30.76	.924	1,523.2	67.64	88.3	.7661	.082	"
5	12.44	12.46	58	293.5	30.74	.928	1,532.2	68.09	88.75	.7672	.082	"
6	12.48	12.50	60	282.5	30.73	.931	1,533.9	67.8	88.88	.7628	.082	"
7	12.52	12.54	48	299.5	30.85	.853	1,347.9	57.5	78.4	.7334	.082	⅞ Gate.
8	12.55	12.57	50	292.5	30.86	.856	1,354.86	58.5	79.02	.723	.082	⅞ "
9	12.58	1	52	283.5	30.88	.859	1,362.12	58.97	79.5	.7243	.082	⅞ "
10	1.05	1.07	38	233	31.18	.684	959.6	35.42	35.41	.6273	.082	½ to ⅝ Gate.
11	1.08	1.10	36	243.5	31.18	.684	959.6	35.06	35.06	.6213	.082	½ to ⅝ "
12	1.11	1.13	34	256.5	31.19	.683	957.42	34.88	34.88	.6194	.082	½ to ⅝ "
13	1.14	1.16	32	270.5	31.21	.678	946.58	34.62	34.62	.6214	.082	½ to ⅝ "

Waste on weir, .082 = 42.15 cu. ft. per minute, deducted from gross amount, to give result in table.

September 25. A. N. Wolff. 24 Inches Diameter.

No. of Test.	Time of Start.	Time of Stop.	Weight Lift'd.	Revolutions per Min.	Head on Wheels.	Head on Weir.	Cubic Feet Discharged per Min.	Horse-Power of Wheel.	Horse-Power of Water.	Percentage of Effect.	Waste on Weir.	Remarks.
	P.M.	P.M.	Lbs.									
1	12.21	12.23	64	266	30.58	.977	1,664.15	68.1	95.97	.7096	.072	Full Gate.
2	12.24	12.26	60	274	30.59	.975	1,659.03	67.95	95.73	.7097	.072	"
3	12.27	12.29	60	287.5	30.58	.9715	1,650.07	69	95.16	.7251	.072	"
4	12.30	12.32	58	297	30.6	.968	1,641.06	68.9	94.79	.7269	.072	"
5	12.33	12.35	56	305	30.6	.96	1,620.65	68.32	93.53	.7305	.072	"
6	12.42	12.44	55	303.5	30.58	.963	1,628.3	66.77	93.9	.7085	.072	"
7	12.45	12.47	57	297.5	30.56	.961	1,623.2	67.83	93.63	.7244	.072	"
8	12.49	12.51	50	276.5	30.79	.842	1,328.05	55.3	76.97	.7184	.072	⅔ Gate.
9	12.52	12.54	44	297.5	30.83	.83	1,300	52.36	75.58	.6927	.072	"
10	1	1.02	30	287.5	31.08	.66	915.37	34.5	53.65	.643	.072	"
11	1.09	1.11	24	272.5	31.4	.572	733.72	26.10	43.45	.6026	.072	⅓ "
12	1.19	1.21	22	282.5	31.45	.56	709.53	24.86	42.08	.5907	.072	⅓ "

Waste on weir, .072 = 34.66 cu. ft. per minute, deducted in column of discharge.

October 15. A. N. Wolff. 24-Inch Wheel. Second Test.

No. of Test.	Time of Start.	Time of Stop.	Weight Lift'd.	Revolutions per Min.	Head on Wheels.	Head on Weir.	Cubic Feet Discharged per Min.	Horse-Power of Wheel.	Horse-Power of Water.	Percentage of Effect.	Waste on Weir.	Remarks.
	P.M.	P.M.	Lbs.									
1	4.34	4.36	54	300	30.18	.913	1,522.52	64.8	86.05	.7478	.04	Full Gate.
2	4.37	4.39	56	291.5	30.16	.918	1,535	65.29	87.31	.7469	.04	"
3	4.41	4.43	58	275	30.12	.923	1,547.5	63.8	87.9	.7258	.04	"
4	4.44	4.46	60	267.5	30.12	.926	1,555	64.2	88.33	.7268	.04	"
5	4.48	4.50	52	312	30.17	.91	1,515	64.9	86.2	.7429	.04	"
6	4.55	4.56	50	320	30.16	.905	1,502.5	64	85.66	.7489	.04	"
7	4.57	4.59	52	307.5	30.14	.909	1,512.5	63.96	85.97	.744	.04	"
8	5.04	5.06	40	300	30.58	.762	1,161.5	48	66.83	.7183	.04	⅔ Gate.
9	5.07	5.08	42	298	30.56	.766	1,170	50	67.43	.7415	.04	"
10	5.10	5.11	30	284	30.83	.644	901.69	34.08	52.43	.6501	.04	"
11	5.17	5.18	24	290	30.89	.582	773.71	27.84	45.18	.6177	.04	⅓ "
12	5.19	5.20	26	271	30.9	.588	785.81	28.18	45.9	.6154	.04	⅓ "

Waste on weir, .040, deducted from discharge.

September 26. John T. Noye & Sons, Buffalo, N. Y. 25 Inches Diameter.

No. of Test.	Time of Start.	Time of Stop.	Weight Lift'd.	Revolutions per Min.	Head on Wheels.	Head on Weir.	Cubic Feet Discharged per Min.	Horse-Power of Wheel.	Horse-Power of Water.	Percentage of Effect.	Waste on Weir.	REMARKS.
	P.M.	P.M.	Lbs.									
1	3.28	3.30	32	285	31.11	.697	995.52	36.48	58.41	.6246	.072	Full Gate.
2	3.32	3.34	34	260	31.1	.698	997.71	36.38	58.52	.6234	.072	"
3	3.35	3.37	30	294	31.1	.691	982.39	35.28	57.62	.6123	.072	"
4	3.38	3.40	28	302.5	31.11	.686	971.44	33.88	56.99	.5944	.072	"
5	3.41	3.43	26	317	31.16	.63	851.96	32.97	50.07	.6585	.072	7/8 Gate.
6	3.45	3.47	26	289	31.24	.62	831.12	30.06	48.97	.6139	.072	3/4 "
7	3.48	3.50	24	300	31.21	.615	820.76	28.8	48.31	.5961	.072	3/4 "
8	3.52	3.53	22	314	31.17	.608	806.33	27.63	47.39	.583	.072	3/4 "
9	3.55	3.57	22	293	31.29	.542	674.23	25.78	39.79	.648	.072	5/8 "
10	4.02	4.04	20	256.5	31.28	.536	660.59	20.52	38.97	.5266	.072	1/2 "
11	4.05	4.07	18	272.5	31.28	.528	647.16	19.62	38.18	.5139	.072	1/2 "
12	4.08	4.10	16	289.5	31.28	.52	631.84	18.52	37.27	.4969	.072	1/2 "
					Continued September 27.							
13	12.24	12.26	30	302	31	.684	967.12	36.24	56.54	.641	.072	Full Gate.
14	12.27	12.29	28	317	30.95	.665	926.09	35.3	54.05	.653	.072	"
15	12.30	12.32	26	325	30.8	.664	923.95	33.6	53.67	.6298	.072	"
16	12.33	12.35	27	320	30.64	.668	932.53	34.56	53.88	.6414	.072	"

Leakage = 34.66 cu. ft. per minute, deducted from discharge.

October 2. Goldie & McCullough, Galt, Province Ontario. 27 Inches Diameter.

No. of Test.	Time of Start.	Time of Stop.	Weight Lift'd.	Revolutions per Min.	Head on Wheels.	Head on Weir.	Cubic Feet Discharged per Min.	Horse-Power of Wheel.	Horse-Power of Water.	Percentage of Effect.	Waste on Weir.	REMARKS.
	P.M.	P.M.	Lbs.									
1	12.18	12.20	52	320	30.25	.945	1,582.55	66.56	90.28	.7373	.072	Full Gate.
2	12.21	12.23	54	316	30.27	.95	1,595.22	68.25	91.06	.7495	.072	"
3	12.27	12.29	58	303.5	30.27	.963	1,628.29	70.41	92.95	.7575	.072	"
4	12.30	12.32	60	301	30.24	.968	1,641.09	72.24	93.59	.7719	.072	"
5	12.33	12.35	62	299	30.2	.972	1,651.34	74.15	94.05	.7884	.072	"
6	12.36	12.38	64	296.5	30.2	.974	1,656.4	75.9	94.34	.8045	.072	"
7	12.39	12.41	66	291	30.18	.982	1,677.03	76.82	95.45	.8048	.072	"
8	12.42	12.44	68	286.5	30.12	.984	1,682.17	77.92	95.55	.8155	.072	"
9	12.45	12.47	70	281.5	30.05	.988	1,692.47	78.82	95.91	.8218	.072	"
10	12.49	12.51	50	280	30.14	.858	1,367.22	56	77.71	.7206	.072	3/4 Gate.
11	12.52	12.54	48	285	30.15	.861	1,374.47	54.72	78.15	.7012	.072	3/4 "
12	12.57	12.59	26	352	30.63	.848	1,343.1	36.61	77.58	.4719	.072	5/8 "
13	1.02	1.04	30	350	30.55	.85	1,347.91	42	77.66	.5408	.072	5/8 "
14	1.07	1.09	30	325	30.65	.782	1,187.25	39	68.62	.5683	.072	1/2 "
15	1.10	1.12	34	312	30.65	.792	1,210.46	42.43	69.97	.6064	.072	1/2 "

Leakage = 34.66 cu. ft. per minute, deducted from discharge.

October 4. John Tyler, Claremont, N. H. 30 Inches Diameter.

No. of Test.	Time of Start.	Time of Stop.	Weight Lift'd.	Revolutions per Min.	Head on Wheels.	Head on Welr.	Cubic Feet Discharged per Min.	Horse-Power of Wheel.	Horse-Power of Water.	Percentage of Effect.	Waste on Welr.	REMARKS.
	P.M.	P.M.	Lbs.									
1	12.21	12.23	74	251	30	.972	1,671.63	74.29	94.57	.7855	.04	Full Gate.
2	12.28	12.30	72	257	30.03	.96	1,640.96	74.02	92.82	.7974	.04	"
3	12.31	12.33	64	278	30.05	.955	1,628.21	71.17	92.27	.7713	.04	"
4	12.34	12.36	62	285.5	30.06	.948	1,610.46	70.8	91.3	.7755	.04	"
5	12.37	12.39	60	288.5	30.1	.945	1,602.87	69.24	90.99	.761	.04	"
6	12.40	12.41	58	294	30.08	.944	1,600.34	68.21	90.78	.7513	.04	"
7	12.46	12.48	64	282.5	30.05	.958	1,635.87	72.32	92.7	.7801	.04	"
8	12.51	12.52	66	277	30.02	.959	1,638.41	73.13	92.76	.7884	.04	"
9	12.54	12.56	58	266.5	30.1	.88	1,441.06	66.47	81.83	.8123	.04	$\frac{7}{8}$ Gate.
10	12.58	12.59	56	261	30.25	.814	1,282.38	58.46	73.16	.7991	.04	$\frac{3}{4}$ "
11	1	1.02	52	246	30.47	.807	1,265.88	51.17	72.74	.7034	.04	$\frac{5}{8}$ "
12	1.03	1.05	48	256.5	30.52	.788	1,221.45	49.25	70.3	.7005	.04	$\frac{1}{2}$ "
13	1.07	1.09	44	247	30.6	.745	1,122.81	43.47	64.79	.6709	.04	$\frac{3}{8}$ "
14	1.10	1.12	40	260	30.65	.73	1,089	41.6	62.95	.6609	.04	$\frac{1}{4}$ "
15	1.14	1.15	36	240	30.8	.622	855.57	34.56	49.7	.6955	.04	$\frac{1}{8}$ "

Waste = 14.352 cu. ft. per minute, deducted from discharge. Friction pulley too high above upper bearing.

October 6. Wm. F. Mosser, Allentown, Pa. 24 Inches Diameter.

No. of Test.	Time of Start.	Time of Stop.	Weight Lift'd.	Revolutions per Min.	Head on Wheels.	Head on Welr.	Cubic Feet Discharged per Min.	Horse-Power of Wheel.	Horse-Power of Water.	Percentage of Effect.	Waste on Welr.	REMARKS.
	P.M.	P.M.	Lbs.									
1	12.26	12.28	42	300	30.56	.763	1,163.8	50.4	67.07	.7514	.04	Full Gate.
2	12.29	12.31	44	289.5	30.58	.772	1,184.46	50.95	68.31	.7459	.04	"
3	12.32	12.34	46	276	30.58	.778	1,198.29	50.78	69.15	.7343	.04	"
4	12.35	12.37	40	312.5	30.6	.77	1,179.86	50	68.09	.7344	.04	"
5	12.38	12.40	38	326.5	30.6	.758	1,151.36	49.63	66.44	.747	.04	"
6	12.41	12.43	36	342	30.62	.752	1,138.69	49.25	66.82	.737	.04	"
7	12.44	12.46	34	356.5	30.65	.744	1,120.55	48.48	64.77	.7485	.04	"
8	12.47	12.49	32	365	30.67	.74	1,111.41	46.72	64.28	.7268	.04	"
9	12.59	1.01	32	310.5	30.9	.666	948.53	39.74	55.27	.719	.04	$\frac{7}{8}$ Gate.
10	1.02	1.04	30	323	30.95	.657	929.27	38.76	54.24	.7146	.04	$\frac{3}{4}$ "
11	1.05	1.07	28	335	31	.648	910.14	37.52	53.21	.7052	.04	$\frac{5}{8}$ "
12	1.08	1.10	26	347	31.02	.644	901.67	36.09	52.75	.6842	.04	$\frac{1}{2}$ "
13	1.11	1.13	24	345	31.03	.636	884.82	34.08	51.78	.6582	.04	$\frac{3}{8}$ "
14	1.18	1.20	24	312	31.1	.595	800.04	29.95	46.92	.6383	.04	$\frac{1}{4}$ "
15	1.21	1.23	22	343	31	.586	781.79	30.18	45.79	.6585	.04	$\frac{1}{8}$ "

Waste = 14.352 cu. ft. per minute, deducted from discharge.

October 10. York Manufacturing Co., York, Pa., Bollinger Wheel. 26¼ Inches Diam.

No. of Test	Time of Start	Time of Stop	Weight Lift'd	Revolutions per Min.	Head on Wheels	Head on Weir	Cubic Feet Discharged per Min.	Horse-Power of Wheel	Horse-Power of Water	Percentage of Effect	Waste on Weir	Remarks
	P.M.	P.M.	Lbs.									
1	12.40	12.42	44	310	30.47	.855	1,380.27	54.56	79.31	.6879	.04	Full Gate.
2	12.43	12.45	48	292.5	30.44	.858	1,367.5	56.16	79.65	.7051	.04	"
3	12.46	12.48	46	300	30.3	.855	1,380	55.2	78.86	.7	.04	"
4	12.50	12.52	44	300	30	.852	1,373	52.8	77.68	.6797	.04	"
5	12.55	12.57	44	290	30.46	.818	1,291.8	51.04	74.2	.6878	.04	¾ Gate.
6	12.58	1	42	300	30.46	.816	1,287	50.4	73.93	.6817	.04	"
7	1.07	1.09	36	290	30.62	.738	1,107	41.76	63.92	.6533	.04	½ "
8	1.10	1.12	34	300	30.61	.737	1,104.74	40.8	63.77	.6398	.04	½ "
9	1.13	1.14	32	306	30.62	.735	1,100.24	39.27	63.53	.6181	.04	½ "
10	1.21	1.23	30	274.5	30.8	.663	942	32.94	54.72	.602	.04	"
11	1.24	1.26	28	290	30.81	.66	935.68	32.48	54.87	.5974	.04	"
12	1.34	1.36	24	263	31	.584	777.81	25.25	45.47	.5553	.04	¼ "
13	1.37	1.39	20	291	31.1	.58	769.51	23.28	45.13	.5158	.04	"

Waste = 14.352 cu. ft. per minute, deducted from discharge.

October 12. Second Bollinger Wheel, York Manufacturing Co. 27 Inches Diam.

No. of Test	Time of Start	Time of Stop	Weight Lift'd	Revolutions per Min.	Head on Wheels	Head on Weir	Cubic Feet Discharged per Min.	Horse-Power of Wheel	Horse-Power of Water	Percentage of Effect	Waste on Weir	Remarks
	P.M.	P.M.	Lbs.									
1	12.35	12.37	34	312	30.59	.759	1,154.6	43.43	66.61	.637	.04	Full Gate.
2	12.38	12.40	36	306	30.55	.763	1,163.8	44.06	66.74	.6576	.04	"
3	12.41	12.43	38	296	30.53	.766	1,170.1	44.99	67.06	.6678	.04	"
4	12.44	12.46	40	287	30.54	.772	1,184.46	45.92	68.22	.6731	.04	"
5	12.47	12.49	42	282.5	30.55	.777	1,198.29	47.46	69.08	.687	.04	"
6	12.50	12.51	44	275	30.53	.782	1,207.54	48.4	69.51	.6963	.04	"
7	12.52	12.53	46	273	30.5	.79	1,226.1	50.23	70.52	.7122	.04	"
8	12.54	12.56	48	270	30.5	.794	1,235.43	51.84	71.05	.7296	.04	"
9	12.57	12.58	50	260	30.48	.798	1,244.76	52	71.55	.7268	.04	"
10	1.01	1.03	52	252.5	30.44	.803	1,256.5	52.52	72.13	.7281	.04	"
11	1.04	1.06	54	246.5	30.42	.804	1,258.84	53.24	72.22	.7372	.04	"
12	1.08	1.09	56	232	30.42	.805	1,261.2	51.97	72.35	.7183	.04	"
13	1.12	1.14	32	318	30.6	.735	1,100.22	40.7	63.49	.641	.04	¾ Gate.
14	1.15	1.17	36	300	30.6	.742	1,116	43.2	64.4	.6708	.04	½ "
15	1.19	1.21	36	280	30.7	.703	1,029	40.32	59.7	.6768	.04	½ "
16	1.23	1.24	34	290	30.72	.7	1,022.3	39.44	59.23	.6659	.04	½ "
17	1.27	1.29	34	268	30.85	.656	927.14	36.45	53.94	.6758	.04	½ "
18	1.30	1.32	32	277	30.85	.654	922.87	35.45	53.69	.6602	.04	¼ "
19	1.40	1.41	24	300	31.03	.6	810.23	28.8	47.41	.6074	.04	¼ "
20	1.42	1.44	26	265	31	.603	816.36	29.64	47.73	.621	.04	¼ "

Waste deducted from discharge as before.

8

October 18.   York Mfg. Co., Center Vent Wheel.   27 Inches Diam.   (Experimental.)

No. of Test	Time of Start	Time of Stop	Weight Lift'd	Revolutions per Min.	Head on Wheels	Head on Weir	Cable Feet Discharged per Min.	Horse-Power of Wheel	Horse-Power of Water	Percentage of Effect	Waste on Weir	REMARKS
	P.M.	P.M.	Lbs.									
1	12.32	12.34	36	265.5	30	.83	1,320.3	38.23	74.7	.5118	.04	Full Gate.
2	12.44	12.46	38	264.5	30.1	.832	1,325.2	40.2	75	.536	.04	"
3	12.47	12.49	40	260.5	30.05	.839	1,341.8	41.68	76.04	.5481	.04	"
4	12.50	12.52	44	255	30.02	.845	1,356.2	44.88	76.79	.5845	.04	"
5	12.53	12.55	46	249	30	.85	1,368	45.81	77.4	.5919	.04	"
6	12.56	12.58	48	244	29.95	.853	1,375.4	46.84	77.61	.6035	.04	"
7	12.59	1	52	240	29.92	.851	1,370.6	49.92	77.19	.6467	.04	"
8	1.01	1.02	56	225	29.86	.86	1,392.87	50.4	78.41	.6428	.04	"
9	1.03	1.05	60	212	29.85	.863	1,399.64	50.88	78.79	.6458	.04	"
10	1.09	1.11	32	264	30.25	.765	1,168.11	33.79	66.64	.5071	.04	¾ Gate.
11	1.15	1.16	30	251	30.48	.667	950.65	30.12	54.61	.5515	.04	½ "
12	1.19	1.21	34	242	30.49	.67	957.11	32.91	55.03	.5982	.04	½ "
13	1.22	1.23	36	233	30.46	.672	961.43	33.55	55.23	.6075	.04	½ "

Waste = 14.352 cu. ft. per minute, deducted from discharge.

October 18.   National Wheel, Bristol, Conn.   25 Inches Diameter.

No. of Test	Time of Start	Time of Stop	Weight Lift'd	Revolutions per Min.	Head on Wheels	Head on Weir	Cable Feet Discharged per Min.	Horse-Power of Wheel	Horse-Power of Water	Percentage of Effect	Waste on Weir	REMARKS
	P.M.	P.M.	Lbs.									
1	3.28	3.30	40	330.5	30.3	.787	1,219.15	52.88	69 66	.759	.04	Full Gate.
2	3.31	3.33	46	312	30.25	.804	1,258.84	57.41	72.05	.7968	.04	"
3	3.34	3.36	50	299	30.24	.814	1,282.38	59.8	73.13	.8177	.04	"
4	3.37	3.39	52	290	30.22	.818	1,291.8	60.32	73.62	.8191	.04	"
5	3.40	3.42	54	287.5	30.2	.823	1,308.4	62.1	74.52	.8334	.04	"
6	3.43	3.45	56	279	30.17	.826	1,310.81	62.49	74.58	.8379	.04	"
7	3.46	3.48	58	265	30.14	.832	1,325.1	61.48	75.32	.8163	.04	"
8	3.49	3.51	60	257.5	30.13	.837	1,337	61.8	75.97	.8135	.04	"
9	3.52	3.54	62	249	30.12	.84	1,344.2	61.75	76.34	.8089	.04	"
10	3.56	3.58	64	241.5	30.11	.844	1,353.77	61.82	76.87	.8042	.04	"
11	4	4 02	66	226.5	30.17	.848	1,363.2	59.8	77.56	.771	.04	"
12	4.11	4.12	32	312	30.67	.7	1,022.4	39.94	59.13	.6754	.04	¾ Gate.
13	4.13	4.14	36	292	30.7	.702	1,026.8	42.04	59.45	.7072	.04	½ "

Waste = 14.352 cu. ft. per minute, deducted from discharge.

October 19. *E. T. Cope & Sons, West Chester, Pa.* 30 *Inches Diameter.*

No. of Test.	Time of Start.	Time of Stop.	Weight Lift'd.	Revolutions per Min.	Head on Wheels.	Head on Weir.	Cubic Feet Discharged per Min.	Horse-Power of Wheel.	Horse-Power of Water.	Percentage of Effect.	Waste on Weir.	Remarks.
	P.M.	P.M.	Lbs.									
1	3.59	4	64	226	30.5	.903	1,497.68	57.86	86.14	.6716	.04	Full Gate.
2	4 03	4.04	60	240	30.55	.894	1,475.44	57.6	84.9	.6784	.04	"
3	4.05	4.06	60	245	30.54	.91	1,515.1	58.8	87.15	.6747	.04	"
4	4.08	4.09	56	272	30.3	.942	1,595.26	60.93	91.91	.6629	.04	"
5	4.10	4.11	58	266	30.26	.943	1,597.8	61.71	92.02	.6706	.04	"
6	4.12	4.13	60	255	30.2	.945	1,602.88	61.2	91.59	.6682	.04	"
7	4.14	4.15	63	254	30	.954	1,625.69	64.01	91.98	.6959	.04	"
8	4.16	4.17	67	240	29.8	.958	1,635.8	64.32	91.93	.6996	.04	"
9	4.19	4.20	70	223	28.5	.954	1,625.69	62.44	87.38	.7146	.04	"
10*	4.37	4.38	64	274	30.15	1.032	1,827.6	70.14	103.92	.675	.04	"
11	4.41	4.42	76	257	29.65	1.035	1,835.3	78.13	102.62	.7613	.04	"
12	4.43	4.44	78	251	29.2	1.04	1,848.74	80.5	101.81	.7907	.04	"
13	4.45	4.46	80	241	28.8	1.04	1,848.74	77.12	100.41	.768	.04	"

Waste = 14.352, deducted from discharge as before.

October 23. *Thomas Tait, Rochester, N. Y.,* "*Centennial Wheel.*" 25 *Inches Diam.*

No. of Test.	Time of Start.	Time of Stop.	Weight Lift'd.	Revolutions per Min.	Head on Wheels.	Head on Weir.	Cubic Feet Discharged per Min.	Horse-Power of Wheel.	Horse-Power of Water.	Percentage of Effect.	Waste on Weir.	Remarks.
	P.M.	P.M.	Lbs.									
1	12.42	12.44	30	331	31.12	.651	916.5	39.72	53.79	.7384	.04	Full Gate.
2	12.37	12.39	32	321.5	31.06	.655	925.01	41.15	54.18	.7595	.04	"
3	12.40	12.41	34	315	31.05	.661	937.82	42.84	54.91	.7801	.04	"
4	12.34	12.36	36	300	31.03	.664	944.24	43.2	55.25	.7818	.04	"
5	12.46	12.48	38	295	31.06	.668	952.84	44.84	55.81	.8033	.04	"
6	12.31	12.33	40	288.5	31	.672	961.43	46.16	56.21	.8212	.04	"
7	12.49	12.51	42	274	31.02	.677	972.24	46.03	56.87	.8093	.04	"
8	12.53	12.54	44	265	31.01	.68	978.75	46.64	57.24	.8148	.04	"
9	12.55	12 56	46	254	31.01	.683	985.25	46.74	57.62	.8112	.04	"
10	12.57	12.58	48	243	31.02	.685	989.59	46.66	57.89	.806	.04	"
11	1.08	1.10	22	322.5	31.32	.539	688.7	28.38	40.68	.6977	.04	$\frac{3}{4}$ Gate.
12	1.04	1.06	24	302.5	31.32	.544	698.42	29.04	41.25	.7039	.04	$\frac{3}{4}$ "
13	1.12	1.13	26	292	31.32	.548	706.25	30.37	41.71	.7282	.04	$\frac{3}{4}$ "
14	1.18	1.20	20	277.5	31.45	.478	573.6	22.2	34.02	.6525	.04	$\frac{1}{2}$ "
15	1.21	1.22	21	265	31.44	.479	575.43	22.26	34.12	.6524	.04	$\frac{1}{2}$ "
16	1.24	1.25	22	258	31.44	.48	577.27	22.7	34.23	.6632	.04	$\frac{1}{2}$ "
17	1.32	1.34	11	268.5	31.57	.37	387.03	11.81	23.04	.5125	.04	$\frac{1}{4}$ "
18	1.30	1.31	12	265	31.47	.373	391.89	12.72	23.26	.5469	.04	$\frac{1}{4}$ "
19	1.35	1.36	13	245	31.57	.377	398.41	12.74	23.72	.5571	.04	$\frac{1}{4}$ "
20	1.43	1.44	7	255	31.6	.305	286.49	7.14	17.07	.4182	.04	$\frac{1}{6}$ "
21	1.39	1.41	8	235	31.59	.309	292.4	7.52	17.42	.4317	.04	$\frac{1}{5}$ "
22	1.47	1.48	5	277	31.61	.301	280.62	5.54	16.73	.3312	.04	$\frac{1}{8}$ "
23	1.45	1.46	6	266	31.61	.303	283.58	6.38	16.9	.3767	.04	$\frac{1}{8}$ "

Waste of 14.352 cu. ft. deducted from discharge as before. The times of making the tests of this wheel are transposed in order to show the increase of weight in regular succession, with the corresponding increase in discharge.

* Stopped for want of steam. These tests were not completed, the steam pumps being unable to supply the necessary water, and the wheel consequently ran very unevenly.

October 31. *Geyelin's Duplex Jonval Turbine. 36 Inches Diameter. R. D. Wood & Co., Philadelphia. Outer Row of Buckets only tested.*

No. of Test.	Time of Start.	Time of Stop.	Weight Lift'd	Revolutions per Min.	Head on Wheels.	Head on Weir.	Cubic Feet Discharged per Min.	Horse-Power of Wheel.	Horse-Power of Water.	Percentage of Effect.	Waste on Weir.	REMARKS.
	A.M.	A.M.	Lbs.									
1	9.17	9.18	34	260	30.1	.7	1,022.38	35.36	58.1	.6086	.04	Full Gate.
2	9.19	9.20	36	255	30.1	.7	1,022.38	36.72	58.1	.632	.04	"
3	9.06	9.08	38	250	30.12	.702	1,026.8	38	58.39	.6508	.04	"
4	9.04	9.05	40	240	30.1	.7	1,022.38	38.4	58.1	.661	.04	"
5	9.23	9.24	42	235	30.3	.703	1,029	39.48	58.86	.6707	.04	"
6	9.25	9.26	44	227	30.3	.703	1,029	39.95	58.86	.6787	.04	"
7	9.27	9.28	46	220	30.3	.703	1,029	40.48	58.86	.6877	.04	"
8	9.29	9.30	48	215	30.3	.705	1,033.4	41.28	59.11	.6983	.04	"
9	9.31	9.32	50	208	30.3	.705	1,033.4	41.6	59.11	.7038	.04	"
10	9.34	9.35	52	200	30.3	.703	1,040.04	41.6	59.49	.6992	.04	"
11	9.37	9.38	54	196	30.28	.708	1,010.04	42.33	59.45	.712	.04	"
12	9.39	9.40	56	190	30.28	.708	1,040.04	42.56	59.45	.716	.04	"
13	9.41	9.42	58	181	30.28	.708	1,040.04	41.99	59.45	.7063	.04	"
14	9.44	9.45	60	170	30.28	.708	1,040.04	40.8	59.45	.6863	.04	"

Wheel bound in case; taken out and cased and repeated trial November 1. Leakage deducted as previously.

	P.M.	P.M.	Lbs.									
15	12.52	12.54	54	205	29.95	.716	1,057.76	44.28	59.81	.7405	.04	Full Gate.
16	12.55	12.56	56	199	29.96	.715	1,055.56	44.57	59.7	.7465	.04	"
17	12.57	12.58	58	191	29.96	.715	1,055.56	44.31	59.7	.7422	.04	"
18	12.59	1	60	186	29.95	.717	1,059.99	44.64	59.98	.7442	.04	"

Wheel still too tight; test stopped.

November 2. Geyelin Duplex Turbine. Both Rows of Buckets.

No. of Test.	Time of Start.	Time of Stop.	Weight Lift'd	Revolutions per Min.	Head on Wheels.	Head on Weir.	Cubic Feet Discharged per Min.	Horse-Power of Wheel.	Horse-Power of Water.	Percentage of Effect.	Waste on Weir.	REMARKS.
	A.M.	A.M.	Lbs.									
1	9.30	9.32	76	223.5	29.53	.912	1,595.28	67.94	88.93	.7639	.04	Full Gate.
2	9.36	9.38	80	217.5	29.55	.942	1,595.28	69.6	88.99	.7821	.04	"
3	9.39	9.41	84	204	29.58	.942	1,595.28	68.54	89.08	.7694	.04	"
4	9.42	9.44	88	195.5	29.58	.94	1,590.23	68.82	88.8	.775	.04	"
5	9.45	9.47	92	185.5	29.52	.938	1,585.2	68.264	88.33	.7711	.04	"
6	9.48	9.49	90	190	29.52	.938	1,585.2	68.4	88.33	.7739	.04	"
7	9.52	9.54	86	197	29.52	.94	1,590.2	67.77	88.62	.7647	.04	"
8	9.55	9.56	82	203	29.56	.943	1,597.8	67.57	89.17	.7581	.04	"
9	10.04	10.06	89	189	29.56	.938	1,585 2	67.284	88.46	.7606	.04	"

Waste = 14 352 cu. ft. per minute, deducted from discharge. This wheel was now withdrawn, the step having worn down one quarter of an inch for want of lubrication, and from the fact that the testing flume was not solid enough to stand the pressure of the water without springing so as to cause the wheel, which was very closely fitted, to bind and wear in its case.

113

November 3. Geyelin Single Jonval Turbine. 36 Inches Diameter. R. D. Wood & Co., Philadelphia.

No. of Test.	Time of Start.	Time of Stop.	Weight Lift'd.	Revolutions per Min.	Head on Wheels.	Head on Weir.	Cubic Feet Discharged per Min.	Horse-Power of Wheel.	Horse-Power of Water.	Percentage of Effect.	Waste on Weir.	REMARKS.
	A.M.	A.M.	Lbs.									
1	9.42	9.44	80	213.5	29.3	.896	1,480.4	68.32	81.89	.8343	.04	Full Gate.
2	9.45	9.47	82	205	29.61	.897	1,482.88	67.24	82.89	.8112	.04	"
3	9.48	9.50	84	197	29.62	.903	1,497.68	66.192	83.75	.7902	.04	"
4	9.51	9.53	86	190.5	29.65	.904	1,500.18	65.53	83.97	.7839	.04	"
5	9.54	9.56	78	211	29.64	.9	1,490.27	65.83	83.39	.7894	.04	"
6	9.57	9.58	78	212	29.62	.9	1,490.27	66.144	83.33	.7937	.04	"
7	9.59	10.01	78	205.5	29.6	.9	1,490.27	64.116	83:28	.7699	.04	"

Waste deducted as previously from discharge, 14.352 cu. ft. This wheel also bound in the step, and was withdrawn for future trials, which were prevented by the close of the Exhibition before the wheel could be refitted. Further trials of the National, Tyler, Cope, and Risdon wheels were also rendered impossible from the same cause.

November 6. Chase Manufacturing Co., Orange, Mass. 24 Inches Diameter.

No. of Test.	Time of Start.	Time of Stop.	Weight Lift'd.	Revolutions per Min.	Head on Wheels.	Head on Weir.	Cubic Feet Discharged per Min.	Horse-Power of Wheel.	Horse-Power of Water.	Percentage of Effect.	Waste on Weir.	REMARKS.
	P.M.	P M.	Lbs.									
1	3.28	3.30	32	399	29.3	.915	1,527.5	51.07	84.69	.603	.04	Full Gate.
2	3.31	3.33	34	380	29.35	.917	1,532.5	51.68	84.92	.6086	.04	"
3	3.34	3.35	34	379	29.52	.918	1,535	51.54	85.54	.6011	.04	"
4	3.36	3.38	36	366	29.7	.921	1,542.5	52.7	86.49	.6093	.04	"
5	3.39	3.41	38	358	29.68	.924	1,550	54.416	87.43	.6224	.04	"
6	3.42	3.44	40	349	29.98	.928	1,560	55.84	88.29	.6324	.04	"
7	3.45	3.47	42	331	30	.927	1,557.5	55.61	88.21	.6304	.04	"
8	3.48	3.49	44	304	29.8	.927	1,557.5	53.5	87.62	.6106	.04	"

Wheel taken out and cased in bearings, and repeated November 7.

9	3.49	3.51	40	365	29.85	.927	1,557.5	58.4	87.77	.6654	.04	Full Gate.
10	3.53	3.54	42	355	29.8	.925	1,552.5	59.64	87.34	.6828	.04	"
11	3.55	3.56	44	332	29.8	.925	1,552.5	58.43	87.34	.669	.04	"
12	3.18	3.20	28	435	29.25	.871	1,419.1	48.72	78.36	.6217	.04	¾ Gate.
13	3.21	3.23	36	363.5	29.35	.866	1,406.9	52.34	77.88	.6714	.04	¾ "
14	3.24	3.26	38	346	29.38	.866	1,406.9	52.59	78.03	.6739	.04	¾ "
15	3.29	3.31	40	325	29.55	.867	1,409.3	52	78.62	.6614	.04	¾ "
16	3.32	3.34	37	360.5	29.62	.867	1,409.3	53.354	78.81	.677	.04	¾ "
17	3.37	3.39	24	372.5	30	.774	1,129.6	35.76	63.97	.559	.04	½ "
18	3.41	3.43	26	350	30.12	.746	1,125.1	36.4	63.97	.5676	.04	½ "
19	3.44	3.46	28	328	30.12	.746	1,125.1	36.74	63.97	.5743	.04	½ "

Waste deducted as previously = 14.352 cu. ft. per minute.

114

November 9. Rodney Hunt, Orange, Mass. 24 Inches Diameter.

No. of Test	Time of Start	Time of Stop	Weight Lift'd	Revolutions per Min.	Head on Wheels	Head on Weir	Cubic Feet Discharged per Min.	Horse-Power of Wheel	Horse-Power of Water	Percentage of Effect	Waste on Weir	Remarks
	A.M.	A.M.	Lbs.									
1	9.27	9.29	58	328	28.96	1.028	1,817.08	76·096	99.35	.766	.04	Full Gate.
2	9.38	9.39	60	317	29.26	1.034	1,832.9	78.616	101.25	.7764	.04	"
	P.M.	P.M.										"
3	12.21	12.23	66	300	29	1.04	1,848.74	79.2	101.21	.7825	.04	
4	12.28	12.30	68	300	29.47	1.045	1,861.98	81.6	103.59	.7877	.04	"
5	12.33	12.34	70	295	29.75	1.047	1,867.31	82.6	104.87	.7876	.04	"
6	12.35	12.36	72	276	29.25	1.038	1,843.45	79.49	101.8	.7809	.04	"
7	12.40	12.42	40	312	30.25	.809	1,270.6	49.92	72.56	.688	.04	½ Gate.
8	12.43	12.45	42	300	30.27	.816	1,284.1	50.4	73.38	.6868	.04	½ "
9	12.46	12.48	44	287.5	30.28	.82	1,296.56	50.6	73.99	.6838	.04	½ "
10	12.50	12.52	54	306	29.8	.966	1,656.2	66.096	93.18	.7094	.04	¾ "
11	12.53	12.55	56	298	29.62	.973	1,674.2	66.75	93.57	.7133	.04	¾ "
12	12.56	12.58	58	289	29.6	.974	1,676.7	67.05	93.7	.7156	.04	¾ "
13	12.59	1	60	278	29.6	.974	1,676.7	66.72	93.7	.7121	.04	¾ "
14	1.03	1.05	36	223	30.45	.748	1,129.63	32.11	64.94	.4945	.04	¼ "
15	1.07	1.08	34	238	30.44	.743	1,118.27	32.37	64.2	.5037	.04	¼ "
16	1.09	1.10	32	254	30.43	.742	1,116	32.51	64.11	.5071	.04	¼ "
17	1.11	1.12	30	266	30.5	.725	1,077.84	31.92	61.98	.5143	.04	¼ "

Waste, 14.352 cu. ft. per minute, deducted as before. It was very difficult to supply this wheel with water, and several tests were rejected in consequence. But a subsequent test of a 54-in. wheel from the same makers, at Passaic, N. J., in January, 1877, gave 141 H. P., equal to .843 per cent., by the water measurement taken by another engineer.

November 10. Stout, Mills & Temple, Dayton, Ohio. 30 Inches Diameter.

No. of Test	Time of Start	Time of Stop	Weight Lift'd	Revolutions per Min.	Head on Wheels	Head on Weir	Cubic Feet Discharged per Min.	Horse-Power of Wheel	Horse-Power of Water	Percentage of Effect	Waste on Weir	Remarks
	P.M.	P.M.	Lbs.									
1	4.42	4.44	64	245.5	30.05	.965	1,653.33	62.848	93.79	.6701	.04	Full Gate.
2	4.45	4.47	66	236.5	29.7	.965	1,653.33	62.436	92.66	.6738	.04	"
3	5.05	5.07	68	236	29.7	.975	1,679.33	64.19	91.09	.6822	.04	"
4	5.08	5.10	70	231	29.55	.98	1,692.38	64.68	94.42	.685	.04	"
5	5.12	5.13	72	223	29.45	.983	1,699.88	64.224	94.51	.6795	.04	"
6	5.19	5.20	74	210	29.05	.973	1,674.2	62.16	91.81	.677	.04	"
7	5.23	5.24	76	201	28.75	.978	1,687	61.1	91.56	.6673	.04	"
8	5.28	5.29	46	259	29.8	.84	1,344.2	47.656	75.62	.6296	.04	½ Gate.
9	5.41	5.42	50	252	29.7	.845	1,356.2	50.4	76.04	.6628	.04	½ "
10	5.43	5.44	54	242	29.48	.854	1,377.85	52.272	78.47	.6661	.04	½ "
11	5.45	5.46	58	230	29.38	.859	1,389.93	53.36	77.09	.6921	.04	½ "
12	6	6.01	34	180	30.95	.548	706.24	24.752	41.26	.5998	.04	¼ "
13	6.02	6.03	34	181	30.95	.547	704.28	24.616	41.15	.5982	.04	¼ "
14	6.04	6.05	28	204	30.95	.535	680.96	22.848	39.79	.5742	.04	Gear on Gate-Shaft loose, &
15	6.06	6.07	24	229	30.97	.527	665.55	21.984	38.91	.5649	.04	Gate closed
16	6.08	6.09	24	224	30.98	.525	661.69	21.5	38.7	.5556	.04	itself grad'ly.

Waste, 14.352 cu. ft. per minute, deducted from discharge. This test was completed by lamplight, at the close of the Exhibition, and some of the latter observations may not be *exactly* correct, but their general correspondence is evidence that they are not far out of the way.

ROVING AND YARN TABLES.

TABLE FOR NUMBERING ROVING BY GRAINS, TROY WEIGHT.

No. of Roving.	Grains per Yard.	Grains per Hank.	No. of Roving.	Grains per Yard.	Grains per Hank.	No. of Roving.	Grains per Yard.	Grains per Hank.	No. of Roving.	Grains per Yard.	Grains per Hank.
.1	83.83	70,000	1.3	6.41	5,384	3½	2.38	2,000	6¾	1.234	1,037
.15	55.56	46,666	1.4	5.95	5,000	3¾	2.22	1,866	7	1.19	1,000
.2	41.66	35,000	1.5	5.55	4,666	4	2.08	1,750	7¼	1.149	965.5
.3	27.77	23,333	1.6	5.2	4,375	4¼	1.96	1,647	7½	1.111	933.3
.4	20.83	17,500	1.7	4.9	4,117	4½	1.85	1,555	7¾	1.075	903.2
.5	16.66	14,000	1.8	4.62	3,888	4¾	1.75	1,473	8	1.041	875
.6	13.88	11,666	1.9	4.38	3,684	5	1.66	1,400	8¼	1.01	848.4
.7	11.9	10,000	2	4.16	3,500	5¼	1.58	1,333	8½	.98	823.5
.8	10.41	8,750	2¼	3.7	3,111	5½	1.51	1,272	8¾	.952	800
.9	9.25	7,777	2½	3.33	2,800	5¾	1.44	1,217	9	.925	777.77
1	8.33	7,000	2¾	3.03	2,545	6	1.38	1,166	..	.....	....
1.1	7.57	6,363	3	2.77	2.333	6¼	1.33	1,120	..	.....	....
1.2	6.94	5,833	3¼	2.56	2,153	6½	1.28	1,076	..	.....	....

TABLE FOR NUMBERING YARN BY GRAINS, TROY WEIGHT.

No. of Yarn.	Grains per Hank.	No. of Yarn.	Grains per Hank.	No. of Yarn.	Grains per Hank.	No. of Yarn.	Grains per Hank.	No. of Yarn.	Grains per Hank.
9	777.77	11¼	622.22	13½	518.51	15¾	444.44	18	388.88
9¼	756.75	11½	608.69	13¾	509.09	16	437.5	18¼	383.56
9½	736.84	11¾	595.74	14	500	16¼	430.76	18½	378.37
9¾	720.51	12	583.33	14¼	491.22	16½	424.24	18¾	373.33
10	700	12¼	571.42	14½	482.75	16¾	417.91	19	368.42
10¼	682.92	12½	560	14¾	474.57	17	411.76	19¼	363.63
10½	666.66	12¾	549.01	15	466.66	17¼	405.79	19½	358.97
10¾	651.16	13	546.15	15¼	459.01	17½	400	19¾	354.43
11	636.36	13¼	526.11	15½	451.61	17¾	394.36	20	350

TABLE FOR NUMBERING YARN BY GRAINS.—(*Continued.*)

No. of Yarn.	Grains per Hank.	No. of Yarn.	Grains per Hank.	No. of Yarn.	Grains per Hank.	No. of Yarn.	Grains per Hank.	No. of Yarn.	Grains per Hank.
20¼	344.44	31½	222.22	42¾	163.74	54	129.62	81	86.4
20½	341.46	31¾	220.47	43	162.79	54¼	129.03	82	85.4
20¾	337.34	32	218.75	43¼	161.84	54½	128.44	83	84.3
21	333.33	32¼	217.05	43½	160.91	54¾	127.85	84	83 3
21¼	329.41	32½	215.38	43¾	160	55	127.27	85	82.4
21½	325.58	32¾	213.74	44	159.69	55¼	126.69	86	81.4
21¾	321.83	33	212.12	44¼	158.19	55½	126.12	87	80.4
22	318.18	33¼	210.52	44½	157.41	55¾	125.56	88	79.5
22¼	314.6	33½	208.95	44¾	156.42	56	125	89	78.6
22½	311.11	33¾	207.4	45	155.55	56¼	124.49	90	77.8
22¾	307.69	34	205.88	45¼	154.69	56½	123.89	91	76.9
23	304.34	34¼	204.3	45½	153.84	56¾	123.34	92	76.1
23¼	301.07	34½	202.89	45¾	152.95	57	122.8	93	75.3
23½	297.87	34¾	201.43	46	152.17	57¼	122.27	94	74.5
23¾	294.73	35	200	46¼	151.3	57½	121.73	95	73.7
24	291.66	35¼	198.58	46½	140.53	57¾	121.21	96	72.9
24¼	288.65	35½	197.32	46¾	149.73	58	120.68	97	72.3
24½	285.71	35¾	195.8	47	148.93	58¼	120.17	98	71.4
24¾	282.82	36	194.44	47¼	148.14	58½	119.65	99	70.7
25	280	36¼	193.1	47½	147.34	58¾	119.14	100	70
25¼	277.22	36½	191.78	47¾	146.59	59	118.47	105	66.7
25½	274.5	36¾	190.47	48	145.83	59¼	118.14	110	63.6
25¾	271.84	37	189.18	48¼	145.07	59½	117.64	115	60.9
26	269.23	37¼	187.91	48½	144.32	59¾	117.15	120	58.3
26¼	266.66	37½	186.66	48¾	143.58	60	116.66	125	56
26½	264.15	37¾	185.42	49	142.85	61	114.8	130	53.8
26¾	261.68	38	184.21	49¼	142.13	62	112.9	135	51.8
27	259.25	38¼	183	49½	141.41	63	111.1	140	50
27¼	256.88	38½	181.81	49¾	140.7	64	109.3	145	48.3
27½	254.54	38¾	180.63	50	140	65	107.7	150	46.7
27¾	252.52	39	179.48	50¼	139.3	66	106.1	155	45.2
28	250	39¼	178.34	50½	138.61	67	104.4	160	43.8
28¼	247.78	39½	177.21	50¾	137.93	68	102.9	165	42.4
28½	245.61	39¾	176.1	51	137.29	69	101.4	170	41.2
28¾	243.46	40	175	51¼	136.58	70	100	175	40
29	241.37	40¼	173.91	51½	135.92	71	98.6	180	38.9
29¼	239.31	40½	172.83	51¾	135.26	72	97.2	185	37.8
29½	237.28	40¾	171.77	52	134.61	73	95.9	190	36.8
29¾	235.29	41	170.73	52¼	133.97	74	94.6	195	35.9
30	233.33	41¼	169.69	52½	133.33	75	93.3	200	35
30¼	231.4	41½	168.67	52¾	132.7	76	92.1	...	....
30½	229.5	41¾	167.66	53	132.07	77	90.9	...	....
30¾	227.64	42	166.66	53¼	131.45	78	89.7	...	....
31	225.8	42¼	165.68	53½	130.84	79	88.6	...	....
31¼	224.08	42½	164.7	53¾	130.23	80	87.5	...	....

ENGLISH TABLE

Showing the quality of Warp Yarn, by the weight, that one seventh of a hank, or eighty turns of a yard-and-a-half reel from one bobbin, will bear before breaking, given in pounds and ounces.

ORDINARY QUALITY.		FAIR QUALITY.		GOOD QUALITY.		EXTRA QUALITY.		SUPER-EXTRA QUALITY.	
No. Yarn.	Breaking Weight.	No. Yarn.	Breaking Weight.	No. Yarn.	Breaking Weight.	No. Yarn.	Breaking Weight.	No. Yarn.	Breaking Weight.
	Lbs. Oz.		Lbs. Oz.		Lbs. Oz.		Lbs. Oz.		Lbs. Oz.
10	115 10	10	120 8	10	125 6	10	130 4	10	135 3
11	102 4	11	104 7	11	106 10	11	108 14	11	111 2
12	96 15	12	99 2	12	100 5	12	103 8	12	105 12
13	91 14	13	93 15	13	96	13	98 2	13	100 4
14	89 12	14	91 12	14	93 13	14	95 14	14	97 15
15	83 12	15	85 10	15	87 8	15	89 7	15	91 6
16	81 11	16	83 8	16	85 6	16	87 4	16	89 2
17	76 14	17	78 10	17	80 6	17	82 2	17	83 14
18	72 10	18	74 4	18	75 14	18	77 8	18	79 3
20	67 14	20	69 6	20	70 14	20	72 7	20	74
22	61 11	22	63 1	22	64 7	22	65 14	22	67 5
24	58 10	24	59 15	24	61 4	24	62 9	24	63 15
26	54 10	26	55 13	26	57 1	26	58 5	26	59 9
28	50 4	28	51 6	28	52 8	28	53 10	28	54 13
30	48 11	30	49 12	30	50 13	30	51 14	30	53
32	45 9	32	46 7	32	47 5	32	48 3	32	49 2
34	44 6	34	45 6	34	46 6	34	47 6	34	48 6
36	41 14	36	42 13	36	43 12	36	44 11	36	45 11
38	39 11	38	40 9	38	41 7	38	42 6	38	43 5
40	38 15	40	39 13	40	40 11	40	41 9	40	42 8
42	37 13	42	38 10	42	39 8	42	40 6	42	41 4
44	35 7	44	36 3	44	37	44	37 13	44	38 10
46	33 13	46	34 9	46	35 5	46	36 1	46	36 14
48	32 3	48	32 14	48	34 0	48	34 5	48	35 1
50	32 2	50	32 13	50	33 8	50	34 4	50	35
55	30 8	55	31 3	55	31 14	55	32 9	55	33 5
60	27 10	60	28 4	60	28 14	60	29 8	60	30 2
65	25 8	65	26 1	65	26 10	65	27 3	65	27 13
70	24 6	70	24 15	70	25 8	70	26 1	70	26 10
75	22 12	75	23 4	75	23 12	75	24 4	75	24 13
80	22	80	22 8	80	23	80	23 18	80	24
85	20 4	85	20 13	85	21 6	85	21 15	85	22 8
90	19 8	90	19 3	90	19 14	90	20 9	90	21 5
95	18 8	95	18 14	95	19 5	95	19 12	95	20 3
100	18 4	100	18 10	100	19	100	19 6	100	19 12
110	15 10	110	16	110	16 5	110	16 11	110	17
120	15 8	120	15 13	120	16 2	120	16 7	120	16 13
130	14 4	130	14 9	130	14 14	130	15 3	130	15 9
140	13 10	140	13 15	140	14 4	140	14 9	140	14 14
150	12 7	150	12 11	150	12 15	150	13 4	150	13 9
160	12 4	160	12 8	160	12 12	160	13	160	13 5
170	11 9	170	11 13	170	12 1	170	12 5	170	12 9
180	10 10	180	10 13	180	11 1	180	11 5	180	11 9
190	10 9	190	10 12	190	11	190	11 4	190	11 8
200	10 4	200	10 7	200	10 11	200	10 15	200	11 3
210	9 13	210	10	210	10 3	210	10 7	210	10 11
220	9 13	220	9 15	220	10 1	220	10 4	220	10 7

ENGLISH TABLE.—(Continued.)

Ordinary Quality.		Fair Quality.		Good Quality.		Extra Quality.		Super-Extra Quality.	
No. Yarn.	Breaking Weight.	No. Yarn.	Breaking Weight.	No. Yarn.	Breaking Weight.	No. Yarn.	Breaking Weight.	No. Yarn.	Breaking Weight.
	Lbs. Oz.		Lbs. Oz.		Lbs. Oz.		Lbs. Oz.		Lbs. Oz.
230	9 3	230	9 6	230	9 9	230	9 12	230	10
240	8 14	240	9 1	240	9 4	240	9 7	240	9 11
250	8 10	250	8 13	250	9	250	9 3	250	9 7
260	8 8	260	8 11	260	8 14	260	9 1	260	9 4
270	8 3	270	8 6	270	8 9	270	8 12	270	8 15
280	8 1	280	8 4	280	8 7	280	8 10	280	8 13
290	7 12	290	7 15	290	8 2	290	8 5	290	8 8
300	7 11	300	7 13	300	8 8	300	8 3	300	8 0
310	7 7	310	7 9	310	7 12	310	7 15	310	8 2
320	7 6	320	7 8	320	7 10	320	7 13	320	8
330	7 2	330	7 4	330	7 17	330	7 10	330	7 13
340	6 15	340	7 1	340	7 3	340	7 6	340	7 9
350	6 14	350	7	350	7 2	350	7 5	350	7 8

ENGLISH TWIST TABLE.

Showing the square root of the numbers or counts from 1 to 100 hanks in the pound, with the twists per inch for different kinds of yarn.

Counts or Numbers.	Square Root.	Frame Warp Twist.	Extra Mule Twist.	Mule Twist.	Filling Twist.	Twist for Doubling.	Hosiery Yarn.
1	1	4.75	4	3.75	3.25	2.75	2.5
2	1.4142	6.71	5.65	5.3	4.6	3.88	3.53
3	1.732	8.22	6.92	6.49	5.62	4.76	4.33
4	2	9.5	8	7.5	6.5	5.5	5
5	2.236	10.62	8.94	8.37	7.25	6.14	5.59
6	2.4494	11.63	9.79	9.18	7.96	6.73	6.12
7	2.6457	12.56	10.58	9.92	8.59	7.27	6.61
8	2.8284	13.43	11.31	10.5	9.19	7.77	7.07
9	3	14.25	12	11.25	9.75	8.25	7.5
10	3.1622	15.02	12.64	11.85	10.27	8.79	7.9
11	3.3166	15.75	13.26	12.43	10.77	9.12	8 29
12	3.4641	16.45	13.85	12.99	11.25	9.52	8.66
13	3.6055	17.12	14.42	13.55	11.71	9.91	9.01
14	3.7416	17.77	14.96	14.03	12.16	10.28	9.35
15	3.8729	18.39	15.49	14.52	12.48	10.65	9.68
16	4	19	16	15	13	11	10
17	4.1231	19.58	16.49	15.46	13.4	11.33	10.3
18	4.2426	20.15	16.97	15.9	13.78	11.66	10.6
19	4.3588	20.7	17.43	16.34	14.16	11.98	10.8
20	4.4721	21.14	17.88	16.77	14.53	12.29	11.18
21	4.5825	21.76	18.33	17.18	14.8	12.6	.....
22	4.6904	22.27	18.76	17.58	15.24	12.8	.....

ENGLISH TWIST TABLE.—(*Continued.*)

Counts or Numbers.	Square Root.	Frame Warp Twist.	Extra Mule Twist.	Mule Twist.	Filling Twist.	Twist for Doubling.	Hosiery Yarn.
23	4.7958	22.78	19.18	17.98	15.58	13.18	.....
24	4.8089	23.26	19.59	18.37	15.92	13.47	.....
25	5	23.75	20	18.75	16.25	13.75	.....
26	5.099	24.22	20.39	19.11	16.57	14.02	.....
27	5.1961	24.63	20.78	19.48	16.88	14.28	.....
28	5.2915	25.13	21.16	19.84	17.19	14.55	.....
29	5.3851	25.57	21.54	20.19	17.49	14.8	.....
30	5.4772	26.01	21.9	20.58	17.8	15.06	.....
31	5.5677	26.44	22.27	19.77	18	15.31	.....
32	5.6568	26.86	22.62	21.21	18.38	15.55	.....
33	5.7745	27.42	23.09	22.64	18.75	15.87	.....
34	5.8309	27.69	23.32	21.86	18.95	16.03	.....
35	5.916	28.1	23.66	22.18	19.22	16.27	.....
36	6	28.5	24	22.5	19.5	16.5	.....
37	6.0827	28.89	24.33	22.81	19.76	16.72	.....
38	6.1644	29.28	24.65	23.11	20.03	16.95	.....
39	6.2449	29.66	24.98	23.41	20.29	17.17	.....
40	6.3245	30.04	25.29	23.71	20.55	17.39	.....
41	6.4031	30.42	25.61	24.01	20.81	17.6	.....
42	6.4807	30.78	25.92	24.3	21.05	17.82	.....
43	6.5574	31.14	26.22	24.59	21.31	18.03	.....
44	6.6332	31.5	26.53	24.87	21.55	18.24	.....
45	6.7082	31.86	26.83	25.15	21.8	18.44	.....
46	6.7823	32.21	27.12	25.43	22.04	18.65	.....
47	6.8556	32.56	27.42	25.7	22.28	18.85	.....
48	6.9282	32.9	27.71	25.98	22.51	18.95	.....
49	7	33.25	28	26.25	22.75	19.25	.....
50	7.071	33.58	28.28	26.51	22.98	19.44	.....
51	7.1414	33.92	28.56	26.78	23.2	19.63	.....
52	7.2111	34.25	28.84	27.04	23.43	19.83	.....
53	7.2807	34.58	29.12	27.3	23.66	20.02	.....
54	7.3484	34.9	29.39	27.55	23.88	20.2	.....
55	7.4161	35.22	29.66	27.81	24.1	20.39	.....
56	7.4833	35.54	29.93	28.06	24.32	20.57	.....
57	7.5498	35.86	30.2	28.31	24.53	20.76	.....
58	7.6157	36.17	30.46	28.55	24.75	20.94	.....
59	7.6811	36.53	30.72	28.8	24.96	21.14	.....
60	7.7459	36.79	30.98	29.04	25.17	21.3	.....
61	7.8102	37.09	31.24	29.28	25.38	21.47	.....
62	7.874	37.4	31.49	29.52	25.59	21.65	.....
63	7.9372	37.7	31.74	29.76	25.79	21.82	.....
64	8	38	32	30	26	22	.....
65	8.0622	38.29	32.24	30.23	26.2	22.17	.....
66	8.124	38.58	32.49	30.46	26.4	22.34	.....
67	8.1853	38.88	32.74	30.69	26.6	22.5	.....
68	8.2462	39.16	32.98	30.92	26.8	22.67	.....
69	8.3066	39.45	33.22	31.14	26.99	22.84	.....
70	8.3666	39.74	33.46	31.37	27.17	23	.....
71	8.4261	40.02	33.7	31.59	27.38	23.17	.....
72	8.4852	40.3	33.94	31.81	27.57	23.33	.....
73	8.544	40.58	34.17	32.03	27.76	23.48	.....
74	8.6023	40.86	34.4	32.25	27.95	23.65	.....
75	8.6602	41.13	34.64	32.47	28.14	23.81	.....

ENGLISH TWIST TABLE.—(Continued.)

Counts or Numbers.	Square Root.	Frame Warp Twist.	Extra Mule Twist.	Mule Twist.	Filling Twist.	Twist for Doubling.	Hosiery Yarn.
76	8.7177	41.4	34.87	32.69	28.33	23.97	.....
77	8.7749	41.68	35.09	32.9	28.51	24.13	.....
78	8.8317	41.95	35.32	33.17	28.6	24.28	.....
79	8.8881	42.21	35.55	33.33	28.88	24.44	.....
80	8.9442	42.48	35.77	33.54	29.06	24.59	.....
81	9	42.75	36	33.75	29.25	24.75	.....
82	9.0553	43.01	36.22	33.95	29.42	24.9	.....
83	9.1104	43.26	36.44	34.16	29.6	25.05	.....
84	9.1651	43.53	36.66	34.36	29.78	25.2	.....
85	9.2195	43.79	36.87	34.57	29.96	25.35	.....
86	9.2736	44.04	37.09	34.77	30.13	25.5	.....
87	9.3273	44.3	37.28	34.97	30.31	25.65	.....
88	9.3808	44.55	37.52	35.17	30.48	25.79	.....
89	9.4339	44.81	37.73	35.37	30.66	25.94	.....
90	9.4868	45.06	37.94	35.47	30.83	26.08	.....
91	9.5393	45.31	38.15	35.77	31	26.23	.....
92	9.5916	45.56	38.36	35.96	31.17	26.37	.....
93	9.6436	45.8	38.57	36.16	31.34	26.51	.....
94	9.6953	46.05	38.78	36.35	31.5	26.66	.....
95	9.7457	46.19	38.98	36.55	31.67	26.8	.....
96	9.7979	46.54	39.19	36.74	31.84	26.94	.....
97	9.8488	46.78	39.39	36.93	32	27.08	.....
98	9.8994	47.02	39 59	37.11	32.17	27.22	.....
99	9.9498	47.26	39.79	37.31	32.36	27.36	.....
100	10	47.5	40	37.5	32.5	27.5	.....

ROVING TWIST TABLE.

Showing twists per inch and the laps per inch on the bobbins, according to the size of the roving.

Hank Roving.	Square Root.	Twist per Inch.	Coils per Inch on Bobbin.	Hank Roving.	Square Root.	Twist per Inch.	Coils per Inch on Bobbin.
½	.7071	.848	6.576	2	1.4142	1.697	13.152
⅝	.791	.949	7.358	2⅛	1.4577	1.749	13.556
¾	.866	1.039	8.052	2¼	1.5	1.8	13.95
⅞	.9354	1.122	8.699	2⅜	1.5411	1.849	14.331
1	1	1.2	9.3	2½	1.5811	1.897	14.704
1⅛	1.0606	1.272	9.863	2⅝	1.6201	1 944	15.067
1¼	1.118	1.341	10.397	2¾	1.6583	1.989	15.422
1⅜	1.1726	1.407	10.805	2⅞	1.6956	2.034	15.768
1½	1.2247	1.469	11.389	3	1.732	2.078	16.107
1⅝	1.2747	1.529	11.849	3⅛	1.7677	2.121	16.439
1¾	1.3228	1.587	12.302	3¼	1.8027	2.163	16.765
1⅞	1.3688	1.643	12.734	3⅜	1.8371	2.204	17.085

ROVING TWIST TABLE.—(*Continued.*)

Hank Roving.	Square Root.	Twist per Inch.	Coils per Inch on Bobbin.	Hank Roving.	Square Root.	Twist per Inch.	Coils per Inch on Bobbin.
3½	1.8708	2.244	17.391	8½	2.9154	3.498	27.113
3⅝	1.9034	2.284	17.701	8⅝	2.9368	3.524	27.312
3¾	1.9364	2.323	18.202	8¾	2.958	3.549	27.509
3⅞	1.9685	2.362	17.313	8⅞	2.979	3.574	27.705
4	2	2 4	18.6	9	3	3.6	27.9
4⅛	2.031	2.437	18.886	9⅛	3.0201	3.624	28.092
4¼	2.0615	2.473	19.165	9¼	3.0413	3.649	28.284
4⅜	2.0918	2.509	19 444	9⅜	3.0618	3.674	28.475
4½	2.1213	2.545	19.723	9½	3.0824	3.698	28.664
4⅝	2.1505	2.58	20	9⅝	3.1024	3.722	28.852
4¾	2.1794	2.615	20.268	9¾	3.1224	3.746	29.039
4⅞	2.2078	2.649	20.533	9⅞	3.1424	3.77	29.224
5	2.236	2.683	20.793	10	3.1622	3.794	29.409
5⅛	2.2638	2.716	21.053	10⅛	3.1815	3.817	29.582
5¼	2.2912	2.749	21.308	10¼	3.2015	3.841	29.774
5⅜	2.3184	2.782	21.561	10⅜	3.221	3.865	29.945
5½	2.3452	2.814	21.81	10½	3.2403	3.888	30.135
5⅝	2.3717	2.846	22.057	10⅝	3.2596	3.911	30.314
5¾	2.3979	2.877	22.3	10¾	3.2788	3.934	30.492
5⅞	2.4238	2.908	22.541	10⅞	3.2975	3.957	30.663
6	2.4494	2.939	22.78	11	3.3166	3.979	30.834
6⅛	2.4748	2.969	23.009	11⅛	3.3354	4.002	31.016
6¼	2.5	3	23.25	11¼	3.3541	4.024	31.193
6⅜	2.5248	3.029	23.48	11⅜	3.3726	4.047	31.365
6½	2.5495	3.059	23.71	11½	3.3911	4.069	31.537
6⅝	2.5739	3.088	23.938	11⅝	3.4095	4.091	31.706
6¾	2.598	3.117	24.161	11¾	3.4278	4.113	31.878
6⅞	2.622	3.146	24.384	11⅞	3.446	4.135	32.047
7	2.6457	3.174	24.605	12	3.4641	4.156	32.216
7⅛	2.6692	3.203	24.823	12⅛	3.482	4.178	32.382
7¼	2.6925	3.231	25.04	12¼	3.5	4.2	32.55
7⅜	2.7156	3.258	25.255	12⅜	3.5176	4.221	32.716
7½	2.7386	3.286	25.468	12½	3.5355	4.242	32.885
7⅝	2.7613	3.313	25.679	12⅝	3.5531	4.263	33.043
7¾	2.7838	3.34	25.89	12¾	3.5707	4.284	33.208
7⅞	2.8062	3.367	26.107	12⅞	3.5881	4 305	33.369
8	2.8284	3.394	26.304	13	3.6055	4.326	33.536
8⅛	2.8504	3.42	26.519	14	3.7416	4.489	34.797
8¼	2.8722	3.446	26.712	15	3.8728	4.647	36.017
8⅜	2.8939	3.472	26.913	...	......	......	.....

RULE BY WHICH TO FIND THE DRAFT OF ANY SPINNING MACHINE.

Write down the number of teeth in all the driving wheels and multiply them together. Then write down the number of teeth in all the wheels that are driven, and multiply them together in like manner. If there is any difference in the diameter of the rollers, multiply the least, or driver's product, by the diameter of the back roller, which is also a driver, and the largest product, or that of the driven wheels, by the diameter of the front roller, which is also driven. Divide the sum of the driven wheels by that of the drivers, and the quotient will be the draft of the machine.

EXAMPLE.

Drivers.	Driven
20	64
18	30
160	1920
20	8
360	2520) 15360 (6.1 draft of frame, nearly.
7	15120
2520	240 remainder = .1, nearly.

TO FIND THE DRAFT ON A MULE.

Suppose the driving pinion on the front roller is 20; stud carrier, 74; change pinion attached to the carrier, 32; this drives the back roller by a wheel of 68. The diameter of the front roller is one inch, and that of the back roller seven eighths of an inch.

RULE.

Multiply the change pinion, 32, by the front-roller pinion, 20, and that product by 7, the diameter of the back roller being seven eighths of an inch. Multiply the number of teeth in the stud carrier, 74, by the number in the roller wheel, 68, and that product by 8, the diameter of the front roller being eight eighths of an inch. Divide the greater number by the less, and the quotient will be the draft of the mule.

EXAMPLE.

Drivers.	Driven.
32	74
20	68
640	592
7 diam. back roller.	444
4480	5032
	8 diam. front roller.
	4480) 40256 (9, Ans.
The draft is nearly 1 into 9.	40320

Rule by which to find the Number of Twists per Inch in the Yarn.

Multiply the number of revolutions of the front roller by its circumference, and divide the number of revolutions of the spindle per minute by that product.

Example.

91 revolutions of front roller per minute
3⅛ inches circumference of roller.
―――
273
13
―――
Inches per minute, 286) 6000 (21 twists to 1 inch, nearly.
572
―――
280
286

To number the yarn produced from a given drawing or sliver: Measure off a convenient number of yards of sliver; multiply this number by extent of drawing on roving and spinning heads; then multiply by 8⅓ and divide by the weight, which will give the number of yarn produced from the given sliver.

Example.

Take 2 yards of sliver weighing 20 grains; 2 × 5, the draw on roving, = 10 × 10, the draw on spinning; 100 × 8⅓ = $\frac{833.3}{20}$ grains = the number, 41.6

To determine the number of hanks or decimal parts of hanks to the pound, for carding, drawing, slubbing, roving, and yarn, according to a given number of yards reeled or measured: Multiply the number of yards by 8⅓ and divide by their weight; the quotient will be the hanks or decimal parts of hanks required.

To determine what weight a given length of drawing, slubbing, roving, or yarn should be to equal a given number of hanks or decimal parts of hanks required: Multiply the given number of yards in length by 8⅓ and divide by the number of hanks or decimal parts of hanks required; the quotient will be the weight of the given length of drawing, roving, or yarn required.

To number the yarn produced by roving: Reel or measure off a convenient number of yards of roving; multiply this number by extent of drawing or spinning heads. This product multiplied by 8⅓ and divided by its weight will give the number of yarn which would be made from the roving.

EXAMPLE.

Suppose 5 yards of roving weigh 20 grains, then 5 × 10 drawing = 50 × 8¼ = $\frac{416.6}{20}$ grains = 20.8, the number.

To change from one number to another on a mule or spinning frame when the draft and roving have both to be altered: Multiply the number of yarn, the yarn being spun, by the hank roving desired, and that product by the number of teeth in the change pinion being used; divide the product thus obtained by the number of yarn desired, multiplied by the hank roving being used. The quotient will show the change pinion required.

To change from one number to another without changing the roving: Multiply the number of teeth in the change pinion in use by the number of yarn being spun. The product, divided by the desired number of yarn, will give the change pinion required.

For the above tables the writer is indebted to Messrs. George Draper & Sons, of Hopedale, Mass.

HISTORICAL SKETCH

OF THE

COMMENCEMENT AND PROGRESS

OF THE

COTTON MANUFACTURE IN THE

UNITED STATES

UP TO

1876.

HISTORICAL SKETCH.

CHAPTER I.

The history of cotton manufacturing in the United States is so inseparably interwoven with that of its progress in Europe, its growth has been so rapid, and its results have exercised such an enormous influence on our national welfare, that it is necessary and advisable, before attempting to describe it, to examine its history in England, where machinery was first applied to this purpose, and note its progress up to the time when the first spindles were set in motion on this side of the Atlantic; and to do this I shall have occasion to quote from the various works of Dr. Andrew Ure on the subject and from Baines's "History of the Cotton Manufacture in Great Britain."

The words "calico," "muslin," and "nankeen" bear testimony to the Asiatic origin of the cotton fabrics bearing those names, which had been imported into Europe long before any attempt was made there to spin the fiber of which they were composed; and any inquiry into the origin or date of the cotton plant and the fabrics produced from it previous to the invention of machinery for the purpose may be dismissed with a short notice as foreign to our subject.

The first record of the introduction of the cotton fiber into England is in the year 1641, in a little treatise on trade, called "Treasure of Traffic," by Lewis Roberts, in which he says: "The town of Manchester buys the linen yarn of the Irish in great quantity, and, weaving it, returns the same again in linen into Ireland to sell. Neither does her industry rest here, for they buy cotton wool in London that comes from Cyprus and Smyrna, and work the same into fustians, vermillions, and dimities, which they return to London, where they are sold; and from thence not seldom are sent into foreign parts, where the first materials may be more easily had for that manufacture."

Were it not for the distinct reference to Cyprus and Smyrna, it would be somewhat doubtful even here if cotton was really the article spun, as the word "cotton" seems to have been indifferently "used for 'coating' in the English works of that day, and denoted a species of woolen stuff made for that purpose." (Ure, "Origin and Progress of Cotton Manufacture," vol. i., p. 30.)

Be this as it may, the amount of cotton used in England was comparatively trifling until the invention of Arkwright in 1768, and it was only in 1774 that it was made lawful by act of Parliament to wear fabrics composed wholly of cotton. Ure says that "the imports of cotton wool from the end of the seventeenth century till the middle of the eighteenth seem, however, to have remained in a stationary condition. In fact, the quantity was only 24,000 or 25,000 lbs. less than 2,000,000 in each of the years 1697, 1701, and 1720. But in 1730 it had fallen to a little more than 1,500,000, and in 1740 it was only 1,660,000. In 1750 it rose to about 3,000,000, and in 1764 it amounted to nearly 4,000,000, betokening the auspicious noonday of the cotton trade of England. The importation of cotton wool was greatly kept in check by the large importation of East Indian cotton goods, which continued with fluctuations during the whole of the eighteenth century, with the exception of a short period toward its close, after the application of the machinery of Arkwright to spin warp, and that of Crompton to spin weft for muslin in general." Ure also says that "almost all the cotton consumed in the British manufactures was obtained from the West Indies and British Guiana prior to the year 1794, with the exception of a little from India and the Levant for the fustian trades, and a still smaller quantity from the Brazils and the Isle of Bourbon for the finer muslin yarns"; and the supply for 1787 is given by him as follows:

British West Indian	6,800,000 lbs.
French and Spanish Colonies	6,000,000 "
Dutch "	1,700,000 "
Portuguese "	2,500,000 "
Isle of Bourbon	100,000 "
Smyrna and Turkey	5,700,000 "
	22,800,000 "

And 26,000,000 lbs. may be considered as the extreme till the appearance in England of cotton from America, which happened, according to Baines, in 1784, when "eight bags of cotton arrived at Liverpool in a vessel from Savannah, and were seized by the customs authorities on the ground that they could not possibly have been the produce of the country whence they were exported." Leaving this matter of the supply of cotton for a while, let us return to the history of its manufacture by machinery.

The first successful patent for drawing cotton by means of rollers revolving at different speeds, which is the whole basis of cotton spinning, was granted to Richard Arkwright in 1769; and in 1771, in connection with Samuel Need, a considerable manufacturing hosier of Nottingham, and Jedediah Strutt, of Derby, the inventor of the frame for making ribbed stockings, he erected the first water spinning mill at Cromford, on the Derwent River in Derbyshire, and in 1775 obtained his second patent, which covered the whole train of operations in a spinning factory.

After Arkwright had at great expense got his mills into successful operation, there arose a number of claimants to different parts of his invention, and it seems probable that crude attempts had been made by other parties at various times, but unsuccessfully, to do that which he succeeded in accomplishing.

At nearly the same time with Arkwright's invention of the "water-frame," as it was called, the spinning jenny was patented by James Hargreaves, differing in principle from Arkwright's process of rollers by having a reciprocating motion and drawing out and twisting the yarn at the same time by the motions of the carriage and spindles, which were, however, separate in the jenny; the spindles being stationary, and the carriage or draw-bar operated with the left hand, regulating the delivery of the roving.

From the "water-twist-frame" of Arkwright and the "jenny" of Hargreaves, in 1770, Samuel Crompton, of Bolton, constructed the "mule" in 1776, which, however, did not come into general use until about 1786, on the abrogation of Arkwright's patent, taking its name of "mule" from its joint parentage.

About the same time an ingenious mechanic of Stockport constructed the "slubbing-billy," a combination of the jenny and the mule, which was used for drawing out the loose "slab" or "slubbing" of wool as delivered from the card, and giving it a partial twist, forming a soft "roving," which was afterward spun into yarn.

In giving Arkwright the credit for the first successful patents for machinery for cotton carding and spinning, it is not my intention to ignore the claims of other parties who had previously made attempts to accomplish the same object by very similar means, and I will therefore briefly mention them. According to Dr. Ure, a patent was granted in 1738 to Lewis Paul, of Birmingham, for "spinning wool and cotton by rollers," but evidence shows the real inventor to have been John Wyatt, of the same town.

I quote from Ure as follows: "An interesting notice of Mr. Wyatt's contrivances for spinning cotton was published by his son, Mr. Charles Wyatt, in the 'Repertory of Arts, Manufactures, and Agriculture' for January, 1818, of which his brother, Mr. John Wyatt, was then editor. The following extracts contain the substance of the communication: 'In the year 1730 or thereabouts, living then at a village near Litchfield, our respected father first conceived the project and carried it into effect ; and in the year 1733, by a model of about two feet square, in a small building near Sutton Coldfield, without a single witness to the performance, was spun the first thread of cotton ever produced without the intervention of the human fingers, he, the inventor, to use his own words, "*being all the time in a pleasing but trembling suspense.*" The wool had been carded in the common way, and was *passed between two cylinders, whence the bobbins drew it by means of the twist.*'" This certainly is not Arkwright's invention, where the "sliver" of cotton is drawn between "pairs of rollers," described by him in this manner in his patent, viz., "Four pairs of rollers, the forms of which are drawn in the margin, which act by tooth and pinion made of brass and steel nuts fixed in two iron plates. That part of the roller which the cotton runs through is covered with wood, the top roller with leather, and the bottom one fluted, which lets the cotton, etc , through it ; by one pair of rollers moving quicker than the other draws it finer for twisting, which is performed by the spindles."

The patent as granted to Paul also claims "a *succession* of other rollers, cylinders, or cones, moving proportionably faster than the first," but unfortunately adds a claim of such manifest absurdity—i. e., "Sometimes these successive rowlers, cylinders, or cones (but not the first) have another rotation besides that which diminishes the thread, viz., that they give it a small degree of twist betwixt each pair, by means of the thread itself passing through the axis and center of that rotation"—as to utterly upset the whole claim. The last paragraph of Paul's patent covers what Wyatt actually did, and what was probably the whole of the invention, viz., "In some other cases only the first pair of rowlers, cillinders, or cones are used, and then the bobbyn, spole, or quill upon which the thread, yarn, or worsted is spun is so contrived as to draw faster than the first rowlers, cillinders, or cones give, and in such proportion as the first mass, rope, or sliver is proposed to be diminished."

In 1748 a patent for carding machinery, in which is described the cylinder· card as first used by hand, was granted to Lewis Paul, and consisted of a cylinder clothed with sheets or fillets, substantially as at the present day. A concave card clothed in the same manner was applied to the under side, and after the cotton was sufficiently carded, by turning the cylinder by hand, the casing was let down, and the cylinder stripped by hand, the rolls obtained in this manner from each sheet being pieced together at the ends to form a continuous roving.

In 1758 a second patent was issued to Paul, from which I quote: "The several rowls or filaments so taken off (the flat cards) must be connected into one entire roll, which being put between *a pair* of rollers or cylinders, is by their turning round *delivered* to the nose of a spindle, in such proportion to the thread made as is proper for the particular occasion. From hence it is delivered to a bobbin, spole, or quill which turns upon the spindle, and which gathers up the thread or yarn as it is spun. The spindle is so contrived as to draw faster than the rollers or cylinders give, in proportion to the length of thread or yarn into which the matter to be spun is proposed to be drawn."

This covers the principal claims to priority of invention in carding and spinning, although the invention of the feeder was claimed by John Lees in 1772, and James Hargreaves, the inventor of the "jenny," claimed the crank and comb for taking the cotton from the card. Thomas Wood also in 1774 claimed to have obtained a perpetual or endless carding by nailing the card fillet on spirally instead of longitudinally; but all these points are covered in Arkwright's patents of 1775.

The machinery of Paul and Wyatt proved a failure, and the mill at Northampton, where it was erected, was dismantled and sold in 1764.

Arkwright's final success led to continual infringements on his patents, and in 1781 a law-suit was the consequence, in which he was beaten on the score of obscurity and defectiveness in his specifications, and a second trial in 1785 confirmed the former decision, and threw his inventions open to the public.

Although Arkwright's first machinery was moved by water power, the invention of the steam engine by Watt in 1769—the same year of Arkwright's first patent—proved of incalculable value to the new manufacture, and in 1785 Messrs. Boulton and Watt erected the first engine for cotton spinning at Papplewick in Nottinghamshire. In 1787 they erected one at Warrington, and three at Nottingham—all for hosiery yarns—and in 1789 one was built for the calico trade of Manchester.

This brings us properly to the end of this chapter and the date of the introduction of the cotton manufacture into America, and it can not better be closed than by the following quotation from Mr. Samuel Batchelder, of Cambridge, to whom the writer is greatly indebted for permission to copy from his valuable little history of the "Introduction of the Cotton Manufacture in the United States," as well as for other information derived from his great experience in manufacturing:

"It is not always easy to decide to whom we ought to award the merit of many inventions, which may have been the study of various ingenious mechanics for years without success; and it happens in relation to cotton machinery, as in other mechanical inventions, that there are conflicting claims to all the most important improvements after they are put in operation. Many may have been engaged for a long time in unsuccessful attempts to accomplish the object, and among them some who have been partially successful, but not so far as to make their schemes of any practical utility. At length some one with better advantages, or better workmanship, or by the application of the same principles with more skill and better judgment, builds a machine which goes into successful operation. In such a case all the unsuccessful schemers rise up and say, 'I tried that principle,' or, 'I put that wheel in operation years ago'; and thus all those who condemn themselves by having made the attempt without success, come before the public and contend for the merit of the more fortunate or more skillful mechanic who has brought the plans to perfection. Something of this kind probably occurred in relation to the invention of Arkwright's spinning machinery. According to the evidence on the trial in relation to his patent in 1785, it would appear that Highs, who" (claimed to have) "invented the spinning jenny in 1763 or 1764, afterward made some experiments or attempts at spinning with rollers, but without succeeding so far as to make it of any practical use. It seems probable that Arkwright became acquainted with the experiments of Highs, and was able, by combination with his own plans, to mature the invention, and put it in successful operation. This, as well as most other important improvements, is the result of successive experiments and failures, until some one who becomes acquainted with the unsuccessful schemes, and has the skill and good judgment to remedy the defects, succeeds in perfecting the invention.

"In 1780 there were twenty water-frame factories, the property of Mr. Arkwright, or of parties who had paid him for permission to use his machinery; and after his patent was made public in 1785, the number increased so rapidly that in 1790 there were one hundred and fifty cotton factories in England and Wales."

CHAPTER II.

In commencing the account of the progress of the cotton manufacture in the United States, the writer must again acknowledge his indebtedness to Mr. Samuel Batchelder, of Cambridge, probably the oldest living cotton manufacturer in the country ;* and to White's "Memoirs of Samuel Slater," published in Philadelphia in 1836, for the greater part of his material relating to the introduction of cotton machinery and the history of its manufacture previous to the foundation of Waltham in 1813. The reader must bear in mind that the factories spoken of so far in England, and which will be mentioned in this country up to the above-mentioned date, bear no comparison to the gigantic structures which strike his eye in all our manufacturing cities and villages to-day : no " Arctic " or " Baltic," " Atlantic " or " Pacific," " Social " or " Harmony " Mills, taking in the cotton at one end, and discharging some completed and beautiful fabric at the other, but small mills of a few hundred or even one or two thousand spindles, simply producing yarn, which was afterward woven by hand in the country farm-houses for miles around into a great variety of coarse " domestic " fabrics ; whence the name applied to the ordinary coarse sheetings and shirtings made by machinery at the present day.

The first record to be found of any action in this country toward introducing machinery for the manufacture of cotton is in the journals of the Legislature of Massachusetts in 1786. I quote from Mr. Batchelder : " On the 25th of October, 1786, Richard Cranch, of the Senate, and Mr. Clarke and Mr. Bowdoin, of the House, were appointed ' to view any new invented machines that are making within this Commonwealth for the purpose of manufacturing sheep's and cotton wool, and report what measures. are proper for the Legislature to take to encourage the same.' This committee reported that ' they had examined those very curious and useful machines made by Robert and Alexander Barr for the purpose of carding and spinning cotton.' And

---

* Since these pages were written, the death of Mr. Batchelder, at Cambridge, Mass., in February, 1879, has closed a long, useful, and valuable life, at the advanced age of over ninety-six years.

in accordance with the further report of the committee, a resolve was passed on the 16th of November, 1786, granting the sum of £200, 'to enable them to complete the said three machines, and also a roping machine, and to construct such other machines as are necessary for the purpose of carding, roping, and spinning of sheep's wool as well as of cotton wool.'"

On the 8th of March, 1787, Messrs. Cranch, of the Senate, and Clarke and Howard, of the House, were appointed a committee to examine the machines now nearly completed by Robert and Alexander Barr, and also to examine and allow their account for the expense of making them, and also to report to the next General Court what gratuity should be allowed them "as a reward for their ingenuity, and an encouragement for their public spirit in making them known to this Commonwealth."

"The report of this committee allowed their account to the sum of £189 12s., including the expense of transporting the machines to and from Boston," from which it is to be inferred that they were exhibited to the Legislature, and on May 2, 1787, a further resolve was passed, discharging the Messrs. Barr from the advance of the £200 aforesaid, and granting them *six tickets* in the land lottery established by an act passed the 14th of November, 1786, as a proper reward. "And it is further resolved, that the said machines be left under the care of the Hon. Hugh Orr, Esq., until the further order of the General Court, and that public notice be given for three weeks successively in Adams and Nourse's Newspaper, that the said machines may be seen and examined at the house of the Hon. Hugh Orr, Esq., in Bridgewater, and that the manner of working them will be there explained to those who may wish to be more particularly informed of their great use and advantage in carrying on the woolen and cotton manufactures. And the said Hon. Hugh Orr, Esq., is hereby requested to explain to such citizens as may apply for the same the principles on which said machines are constructed, and the advantages arising from their use, both by verbal explanations and by letting them see the machines at work. And it is further resolved, that the said Hon. Hugh Orr, Esq., be, and he hereby is, permitted and allowed to make use of the said machines during the whole time of his having the care of them, as some recompense for his own time and trouble in shewing them and explaining their use to the citizens of this commonwealth at large."

Mr. Batchelder quotes from Judge Mitchell's "History of Bridgewater" the following notice of the above-mentioned Hugh Orr, Esq.: "Hugh Orr was born at Lochwinnoch, in Scotland, January 2, 1715, and came to America June 17, 1740, and settled at Bridgewater, where he died December 6, 1798. He was engaged there before the

Revolution in the manufacture of fire-arms, and at the commencement of that war made the first cannon that were made in this country by boring from the solid casting. He is said to have invited Robert and Alexander Barr, both Scotchmen, brothers, in order to construct at his works in East Bridgewater machinery for carding, roving, and spinning cotton."

In the "Memoirs of Samuel Slater" is given in full the petition of Thomas Somers, said to have been a midshipman in the English Navy, which was presented to the Legislature of Massachusetts about the same time, and set forth—"That in the fall of the year 1785, the tradesmen and manufacturers of Baltimore in Maryland, having formed themselves into an association, in order to apply to the Legislature in behalf of American manufactures, being stimulated thereto by a circular letter received from a committee of the tradesmen and manufacturers of the town of Boston : your petitioner, then residing at Baltimore (having been formerly brought up to the cotton manufactory, and willing to contribute what lay in his power to introduce said manufacture in America), did at his own risk and expense go to England in order to procure the machines for carding and spinning cotton. That, after much difficulty, your petitioner found that he could only take descriptions and models of said engines ; with which he returned to Baltimore last summer. Soon after his arrival he found that they were very dilatory about encouraging the matter, and with the advice of some friends he resolved to try what might be done in Boston. That, on his way to Boston, the sloop was driven ashore by the late storms on Cape Cod, by which misfortune your petitioner lost almost one half the small property he had to subsist on till he could get into business. Your petitioner therefore humbly prays for such assistance to begin the manufactory as shall seem most agreeable to your honors," etc., etc. "N. B. Your petitioner is perfect master of the weaving in the speediest manner, and of adapting to advantage the different kinds of yarn for Marseilles quilting, dimity, muslins plain, striped, or checked, calico, cotton and linen jeans, jeanettes, handkerchiefs, checks, drabs, and many other kinds in that line, and understands the management of cotton, and how the spinning should be performed."

On the 8th of March, 1787, the Legislature of Massachusetts passed a resolution appropriating £20 for the purpose of giving Somers "an opportunity to give specimens of his ability to perfect the manufactures, as set forth in his petition" ; to be deposited in the "hands of Hon. Hugh Orr, who shall be a committee to superintend the application of the same."

The same Somers afterward appears in connection with the factory at Beverly, Mass., which was also projected in 1787 by Messrs. John

Cabot and others, and which appears, according to Mr. Batchelder, to have been the first one to produce yarn by machinery in the United States, as it was undoubtedly in operation some time before 1789. There seems to be no doubt that the machinery built at Bridgewater was the first on the Arkwright plans, but it does not seem to have been put in practical operation ; and the probability is that the spinning at Beverly was done on the Hargreaves "Jenny."

Finding the construction of the machinery very difficult and expensive and the prospects discouraging, the proprietors applied to the Legislature for aid, which was granted by the following resolve, February 17, 1789 : "Be it resolved, That there be granted, and there is hereby granted accordingly, and conveyed to John Cabot, Joshua Fisher, Henry Higginson, Moses Brown, George Cabot, Andrew Cabot, Israel Thorndike, Isaac Chapman, and Deborah Cabot, they being members of the said corporation, the value of five hundred pounds, lawful money in specie, to be paid in the eastern lands, the property of this commonwealth," etc., the lands being assigned in different proportions, from one fortieth part to ten fortieths parts, to the above named proprietors ; conditional, however, on the manufacture within the next seven years of "a quantity of not less than 50,000 yards of cotton and linen piece goods, of a quality usually imported into this commonwealth," of which a true record was to be kept and a fair copy lodged in the office of the Secretary of State, verified by the oath of at least two of the proprietors ; and failing which the lands were to revert to the commonwealth, unless the said corporation should pay to the Treasurer of the Commonwealth £500 in gold or silver within eight years from the passage of the resolve. At the same session of the Legislature an act was passed incorporating the aforesaid parties, including Thomas Somers, as the "Beverly Manufacturing Company," and authorizing them to hold personal property to the amount of £30,000 and real estate to the amount of £10,000.

In June, 1790, the same parties presented another petition, signed by John Cabot and Joshua Fisher, managers, representing "that they had expended about £4,000, and that the present value of their stock was not equal to £2,000, and that a further very considerable advancement is absolutely necessary ; that the intended aid by a grant of land made by a former Legislature has not in any degree answered the purpose of it ; and pray that in lieu of that grant some real and ready assistance may be afforded them."

"The petitioners state, as one of the public advantages to be derived from the manufacture of cotton, that the raw material is procured from the West Indies, in exchange for fish, 'the most valuable export in possession of the State.' They also mention the extraordinary cost of machines, intricate and difficult in their construction, without any model in the country, and instance a carding machine that cost $1,100." The Legislature voted them "a grant of £1,000, to be raised in a lottery, on condition that they give bonds that the money be actually appropriated in such a way as will most effectually promote the manufacturing of cotton piece goods in this commonwealth." Mr. Batchelder then quotes from Washington's diary as follows:

"Friday, October 30, 1789. After passing Beverly two miles, we came to a cotton manufactory, which seems to be carrying on with spirit by the Cabots (principally). In this manufactory they have the new invented carding and spinning machines. One of the first supplies the work, and four of the latter, one of which spins 84 threads at a time by one person. The cotton is prepared for these machines by being first (lightly) drawn to a thread on the common wheel. There is also another machine for doubling and twisting the thread for particular cloths; this also does many at a time. For winding the cotton from the spindles and preparing it for the warp there is a reel, which expedites the work greatly." "A number of looms (15 or 16) were at work with spring shuttles, which do more than double work. In short, the whole seemed perfect and the cotton stuffs which they turn out excellent of their kind—warp and filling both of cotton."

This factory was built of brick, and continued in operation for several years, and was driven by horse-power, and appears to have been, by the above extract, indisputably the earliest enterprise carried into execution in this country.

A great deal of interest was also manifested in Philadelphia at the same period on the subject of manufactures, and Tench Coxe, who was Assistant Secretary of the Treasury under Hamilton, delivered an address August 9, 1787, to an assembly of the friends of American manufactures, convened for the purpose of establishing a "*Society for the Encouragement of the Useful Arts.*"

Mr. Samuel Wetherill, Jr., as chairman, signed a report of the managers of the society in August, 1788, by which it appears that the amount of cash received from the contributors on the 23d of August was £1,327 10s. 6d.; that they had purchased a quantity of flax, and employed between two and three hundred women in spinning linen yarn, and also engaged workmen to make a carding engine, and four jennies, of 40, 44, 60, and 80 spindles, for spinning cotton; that as soon as the season would permit the house to be fitted up, they were set to work, but, owing to various delays and obstructions thrown in their way *by foreign agents*, it was the 12th of April, 1788, before they began to weave, and on the 23d of August, 1788, they had made 11,367 yards of various kinds of cotton and linen goods.

Mr. Wetherill had been engaged in manufacturing for some years, as appears by his advertisement in the "Pennsylvania Gazette" of April 3, 1782, of :

"Philadelphia Manufactures, suitable for every season of the year, viz.: Jeans, Fustians, Everlastings, Coatings, etc., to be sold by the subscriber at his dwelling-house and manufactory, in South Alley, between Market Street and Arch Street, and between Fifth and Sixth Streets, on Hudson's Square. SAMUEL WETHERILL."

The manufacturers of Rhode Island were also turning their attention to the new machinery at this time, as will be seen by the following account, furnished by William Anthony, which I copy from the memoir of Slater :

"About the year 1788 Daniel Anthony, Andrew Dexter, and Lewis Peck, all of Providence, entered into an agreement to make what was then called 'homespun cloth.' The idea at first was to spin by hand, and make jeans with linen warp and cotton filling, but, hearing that Mr. Orr, of Bridgewater, Mass., had imported some models of machinery from England for the purpose of spinning cotton, it was agreed that Daniel Anthony should go to Bridgewater and get a draught of the model of said machine ; he, in company with John Reynolds, of East Greenwich, who had been doing something in the manufacturing of wool, went to Bridgewater and found the model of the machine spoken of in possession of Mr. Orr, but not in operation. It was not the intention of Mr. Orr to operate it, but he only kept it for the inspection of those who might have an inclination to take draughts. The model of the machine was very imperfect, and was said to be taken from one of the first built in England. A draught of the machine was accordingly taken, and laid aside after a while. They then proceeded to build a machine of a different construction called a jenny ; I understood that a model of this machine was brought from England into Beverly, Mass., by a man of the name of Somers. This jenny had 28 spindles ; the woodwork was built by Richard Anthony ; the spindles and brasswork were made by Daniel Jackson, an ingenious coppersmith of Providence. This jenny was finished in 1789. It was first set up in a private house, and afterward removed to the market-house chamber in Providence, and operated there. Joshua Lindly, of Providence, was then engaged to build a carding machine, for carding the cotton agreeably to the draught presented, also obtained from Beverly. This machine was something similar to the one now used for carding wool, the cotton being taken off the machine in rolls, and roped by hand; after some delay this machine was finished. They then proceeded to build a spinning frame after the draught obtained at Bridgewater. This machine was something similar to the water frame now in use, but very imperfect; it consisted of 8 heads of 4 spindles each, being 32 spindles in all, and was operated by a crank turned by hand. The first head was made by John Baily, an ingenious clock-maker of Pembroke, Mass.; the other seven heads were made by Daniel Jackson, of Providence. The woodwork was made by Joshua Lindly. In 1788 Joseph Alexander and James McKennis, natives of Scotland, arrived in Providence, both being weavers and understanding the use of the fly-shuttle; they were engaged to weave corduroy, Mr. Alexander to weave a piece in Providence, and Mr. McKennis went to East Greenwich to work there. A loom was accordingly built after the direction of Mr. Alexander, and put in operation in the market-house chamber; this was the first fly-shuttle ever used in Rhode Island. A piece of corduroy was then woven, the warp being linen and the filling cotton, but, as there was no person to be found who could cut the corduroy and raise the pile which makes the ribs on the face of the cloth, and give it the finish, it was thought best to abandon that kind of cloth. Mr. Alexander went to Philadelphia. Mr. McKennis continued to work in Greenwich for some years. This appears to be the beginning of the jenny-spinning in Rhode Island, and undoubtedly originated with the above company.

"The spinning frame (the one attempted from the State's model), after being tried some time in Providence, was carried to Pawtucket and attached to a wheel propelled by water. The work of the machine was too laborious to be done by hand, and the machine was too imperfect to be turned by water. Soon after this the machine was sold to Mr. Moses Brown, of Providence, but, as all the carding and roping was done by hand, it was very imperfect, and but little could be done. This was the situation of cotton manufacturing in Rhode Island when Mr. Samuel Slater arrived in this country; then all this imperfect machinery was thrown aside, and machinery more perfect built under his direction."

This statement is confirmed by Joseph Anthony, the son of the Daniel Anthony above mentioned. The Mr. Moses Brown to whom the water frame was sold was a partner of the firm of Almy & Brown, who were about commencing the business of what was strictly cotton *manufacturing*, the yarn being spun and the cloth woven by manual labor.

A statement of their production from the commencement, June 11, 1789, to January 1, 1791, shows:

Corduroy	45 pieces,	1,090 yards,	sold from	3s. 6d. to 4s.	per yard.
Denims, royal ribs, etc.	25 "	558 "	"	3s. to 4s.	"
Cottonets	13 "	325 "	"	2s. 6d. to 3s.	"
Jeans	79 "	1,897 "	"	2s. to 2s. 6d.	"
Fustians	26 "	687 "	"	1s. 8d. to 2s.	"
Total	189	4,556			

With this summary of the progress of the cotton manufacture and its condition in 1789, I will close this chapter, and in the next will introduce Samuel Slater, whose arrival in the United States with the necessary information marks the era of positive and decided progress.

CHAPTER III.

The following brief account of himself, found among Mr. Slater's papers, forms a fitting opening to this chapter :

"Samuel Slater was born in the town of Belper, in the county of Derby, June 19, 1768. On June 28, 1782, being about fourteen years of age, he went to live with Jedediah Strutt, Esq., in Milford, near Belper (the inventor of the Derby ribbed stocking machine and several years a partner of Sir Richard Arkwright in the spinning business), as a clerk, who was then building a large factory at Milford, where said Slater continued until August, 1789. During four or five of the last years his time was solely devoted to the factory as general overseer, both as respected making machinery and the manufacturing department. On the first day of September, 1789, he took his departure from Derbyshire for London, and on the 13th he sailed for New York, where he arrived in November, after a passage of sixty-six days. He left New York in January, 1790, for Providence, and there made an arrangement with Messrs. Almy & Brown to commence preparation for spinning cotton at Pawtucket.

"On the 18th day of the same month the venerable Moses Brown took him out to Pawtucket, where he commenced making the machinery, principally with his own hands, and on the 20th day of December following he started three cards, drawing and roving, and 72 spindles, which were worked by an old fulling-mill water-wheel in a clothier's building, in which they continued spinning about twenty months ; at the expiration of which time they had several thousand pounds of yarn on hand, notwithstanding every exertion was made to weave it up and sell it. Early in the year 1793 Almy, Brown & Slater built a small factory in that village (known and called to this day the old factory), in which they set in motion July 12 the *preparation* and 72 spindles, and slowly added to that number as the sales of the yarn appeared more promising, which induced said Slater to be concerned in erecting a new mill, and to increase the machinery in the old mill."

Slater's motive for leaving England is said to have been his observing in a Philadelphia paper an advertisement of a reward offered by a society for a machine to make cotton rollers, etc. This convinced him that there was an opportunity to turn his knowledge to account in this country, and, fearing that the cotton business "*would be overdone*" in England, he resolved to emigrate. As the laws of England prohibiting the exportation of machinery were very severe, he took no patterns or drawings of any kind with him, trusting solely to his excellent memory, and relying for an introduction on his indenture as an apprentice to Jedediah Strutt. Landing in New York, he was introduced to the "New York Manufacturing Company," and entered their employment; but, not liking the prospects which were opened to him, and hearing through the captain of one of the Providence packets of Moses Brown, he wrote to him, with the result above stated. The old machinery, which, as has been related, Mr. Brown had purchased, was first shown him, but condemned by him at once as unsatisfactory, and he immediately commenced building a new set.

With the introduction of the improved machinery by Slater, the manufacture of cotton in the United States may be said to have fairly commenced, and some of the first yarn, said to have been as fine as No. 40, with some of the first cloth made from the same warp, was sent to the Secretary of the Treasury October 15, 1791.

He, however, found great difficulty in procuring proper mechanical assistance to build the machinery from his instructions, and his greatest perplexity was in making the cards, for which purpose he employed Pliny Earle, of Worcester, who had never before made machine cards, but finally succeeded in accomplishing the desired result; and the demand for cards which was created by the success of the new manufacture resulted in the invention of the card-setting machine by Amos Whittemore, of Cambridge, in 1797, and its subsequent introduction in England in 1799. This, however, must be considered as only the second great American invention relating to the manufacture of cotton, the first having been the cotton gin, which was the invention of Eli Whitney in 1793.

This leads us to the consideration of another branch of the subject, viz., the adequate supply of the raw material in proper condition for manufacture.

The First Provincial Congress in South Carolina, held in January, 1775, recommended to the inhabitants "to raise cotton," yet very little practical attention was paid to their recommendation. A small quantity only was raised for domestic manufactures. Georgia took the lead in this culture, and the introduction of the new machines and the consequent demand greatly promoted it. We find in "Baines's History of Cotton Manufacture" the export from the United States in 1791 given as 189,316 lbs.; and in 1792 as 138,328 lbs.; in 1793, 487,600;

and in 1794, after the invention of the gin, it rose to 1,601,700 lbs., and thenceforward the increase was constant and rapid.

Cotton had been produced for a long time in small quantities in several of the Southern States, and the following extract from a pamphlet by Dr. G. Emerson, of Philadelphia, entitled " Cotton in the Middle States," published in 1862, which I copy from Mr. Batchelder, is worthy of introduction in this place :

" Long before the Southern States took up its regular culture, cotton was raised on the eastern shore of Maryland, lower counties of Delaware, and other places in the Middle States. As early as 1736, and for some time after, it was chiefly regarded as an ornamental plant, and confined to gardens ; but it soon became appreciated for its useful qualities, and was brought under regular cultivation. This culture, though comparatively limited in those places, has never been entirely abandoned up to the present day. I have myself seen many families who came from Sussex County, Delaware, to reside in the adjoining county of Kent, wearing clothes made of cotton of their own raising, spinning, and weaving. The culture of cotton in this section of our country gradually diminished, in consequence of the vast area over which the plant was extended in more southern States. In competition with these, our more northern farmers found they possessed superior advantages for raising other field-crops, from which they derived greater profits."

" Limited as has been the culture of cotton on the peninsula between the Delaware and Chesapeake Bays, it has furnished a demonstration of the highest importance to our country. In proof of this it may be stated that at the close of the Revolution a convention was held at Annapolis, in 1786, to consider what means could be best resorted to for the purpose of remedying the embarrassment of the country, then so much exhausted in its finances.

"The late President Madison, a member of this convention from Virginia, there expressed it as his opinion, '*that, from the results of cotton raising in Talbot County, Maryland, and numerous other proofs furnished in Virginia, there was no reason to doubt that the United States would one day become a great cotton-producing country!*' It would hence appear that the first culture of cotton in the United States worthy of notice was made in the peninsula between the Delaware and Chesapeake Bays, from whence it crossed into Western Maryland and Virginia, and so went southward."

It would, however, appear, as shown by the following letter, that the first cotton received by Messrs. Slater & Brown was so imperfectly cleaned as to be of small comparative value, and we find that, when they first began to spin, they used Cayenne and Surinam cotton, but after a few years they began to mix about one third of Southern cotton, and this yarn was designated as second quality and sold at a price accordingly.

On the 19th of April, 1791, Moses Brown writes to the proprietors of the Beverly factory as follows :

"I have for some time thought of addressing the Beverly manufacturers on the subject of an application to Congress for some encouragement to the cotton manufacture by an additional duty on the cotton goods imported, and the applying such duty as a bounty, partly for *raising and saving cotton in the Southern States, of a quality and cleanness suitable to be wrought by machines*, and partly as a bounty on cotton goods of the kind manufactured in the United States."

On the 15th of November, 1791, Mr. Brown writes to J. S. Dexter on the same subject as follows :

"PROVIDENCE, *November* 15, 1791.

"When it is considered that cotton, the raw material, may be raised in the United States, it shows that legislative attention should be paid to this subject. The cotton raised at present in the Southern States is as imperfect as our manufactured goods. This, I presume, is owing to the promiscuous gathering and saving of the article from the pods in which it grows, some of which, like fruit on a tree, are fair and full grown, while others are not. In the picking of these, and in taking the cotton out of the pods, care should be taken that it be kept separate, and the thin membrane which lines the pod, and sometimes comes off with the cotton, should be separated, and the clean, full grown preserved to work on the machines ; the other will answer to work by hand. But, as the cotton must be clean before it works well on the card, the present production, in the mixed manner in which it is brought to market, does not answer a good purpose. The unripe, short, and dirty part, being enveloped with that which would be good if separated properly at first, so spoils the whole as to discourage the use of it in the machines, and obliges the manufacturer to have his supply from the West Indies, under the charge of the impost, rather than work our own production—a circumstance truly mortifying to those who, from motives of promoting the produce and manufactures of our own country, as well as from interest, have been at much expense and trouble to promote so desirable an object. I therefore beg leave to suggest the idea of some encouragement to the raising and saving of cotton, clean and fit for the manufacturers."

The relief from these difficulties was soon provided by the ingenuity of Eli Whitney.

Born at Westboro, Worcester County, Mass., December 8, 1763, he developed indications of mechanical genius at a very early age. When twelve years old he "made a fiddle," in his sister's words, and after that he was often employed to repair violins. By his own personal exertions he prepared himself for Yale College, which he entered in May, 1780, and through which he passed with little expense to his father. On one occasion he repaired the philosophical apparatus belonging to the college, to the great satisfaction of the Faculty.

Soon after taking his degree in the autumn of 1792, he formed an engagement with a gentleman of Georgia to reside in his family as a tutor, and on his way thither was so fortunate as to fall into the company of the widow of General Greene, who with her family was returning to Savannah after spending a summer at the North. On arriving in Georgia, he found that the gentleman who had engaged him had employed another tutor, leaving him entirely without resources or friends, except those he had made in the family of General Greene. The interest he had excited in them, however, led to a kind invitation from Mrs. Greene to make her house his home, and there pursue his studies, which he accepted, and commenced the study of law under her hospitable roof.

Turning his mechanical ingenuity to account, he soon made for Mrs. Greene a tambour frame; and not long after this incident a party of gentlemen, principally officers who had served under General Greene in the Revolutionary war, came from Augusta and the upper country on a visit to the family.

The conversation turned one day on the state of agriculture among them, and great regret was expressed that there were no means of cleaning the green-seed cotton or separating it from its seed, since all the lands which were unsuitable for the culture of rice would yield large crops of cotton.

During this conversation Mrs. Greene said, "Gentlemen, apply to my young friend Mr. Whitney; he can make anything." She then led the company to another room, and showed them the tambour frame which he had made, and also a number of toys which he had made or repaired for the children, and then introduced them to Mr. Whitney himself.

Mr. Whitney disclaimed all pretension to mechanical genius, and said that he had never seen either cotton or cotton seed in his life; but a new turn was given to his views, and he went to Savannah, and searched the warehouses and boats till he found a small parcel of cotton in the seed. This he took home with him, and commenced his experiments with such rude tools as he could find, even drawing his own wire, of which the teeth of the first gins were composed.

The ensuing winter saw the new machine completed, and Mrs. Greene invited to her house a number of gentlemen from different parts of the State to witness the new invention. They saw with astonishment and delight that more cotton could be separated from the seed with it in one day, by the labor of a single hand, than could be done in the former manner in many months.

Phineas Miller, Esq., a native of Connecticut and a graduate of Yale College, who married the widow of General Greene, contributed much to the success of the undertaking. He provided the funds to carry out the enterprise, and the parties agreed to share the profits and emoluments resulting by an instrument bearing date May 27, 1793. Immediately after this they commenced business under the name of Miller & Whitney. On the 25th of June, 1793, Whitney presented his petition for a patent to Thomas Jefferson, then Secretary of State, and on the 20th of October in the same year he made oath to his invention before the Notary Public of the city of New Haven.

Of his long and tedious struggles with the horde who grasped at his invention, without any remuneration to him, of the almost endless litigations and disappointments which followed, I have neither the time nor the space to speak here; it is sufficient for our present purpose that the invention was made, the supply of the raw material to the Northern manufacturers assured, to say nothing of the wants of Europe, and the destiny of the Southern States of the Union fixed for a century at least. Next on the roll of inventors to Arkwright, in point of time as well as importance in the history of the cotton manufacture, stands Eli Whitney, the first American who is distinguished in that connection. Next in order to Whitney comes Whittemore, already mentioned, whose machine for setting card clothing is often selected as an example of the perfection of mechanical automatism. This was soon adopted by Pliny Earle, whose nephews still carry on the business of making card clothing in Worcester, under the firm name of Timothy K. Earle & Co. This closes the period of distinct invention for the century, and, although many small modifications and improvements may have been made, we shall find little to note except the growth of the now established business of "cotton spinning" until the War of 1812, the introduction of the power loom, and the building of the first mill at Waltham for combining all the processes of making cloth under one roof. Meanwhile we will devote another chapter to notices of the extension of the business, which was very rapid, and which spread to various parts of the country during the intervening period.

CHAPTER IV.

In 1798 Samuel Slater entered into partnership with Oziel Wilkinson, Timothy Green, and William Wilkinson, the two latter as well as himself having married daughters of Oziel Wilkinson. He built the second mill on the east side of Pawtucket River, called the "White Mill," in what was then the town of Rehoboth, within the limits of Massachusetts, and an act was passed by the Massachusetts Legislature in 1799 exempting the said mill, together with the materials and stock, from taxation for seven years from April 1, 1800.

The firm was known as Samuel Slater & Co., he holding one half the stock.

"Until this time" (according to Mr. Batchelder) "the business had been confined to Slater and his associates, but soon after this it is stated that several of his men who had become acquainted with the construction of his machinery left his employment, and commenced the erection of mills for themselves or other parties. Mr. Benjamin S. Wolcott was employed by Mr. Slater in the construction of his first mill. After acquiring sufficient knowledge of the business, he united with Rufus and Elisha Waterman for the purpose of erecting a cotton factory in Cumberland about 1801. The machinery was afterward removed to Central Falls, a short distance above Pawtucket, and a new company formed, with the addition of Mr. Stephen Jenks.

Another of his workmen, by the name of Robbins, commenced a mill in New Ipswich, which was put in operation in 1804; being the first cotton mill built in New Hampshire.

B. S. Wolcott, Jr., was employed in one of the early mills at Pawtucket; a second one, known as the "Yellow Mill," having been built in 1805, under an act exempting it from all taxes for five years; and with the assistance of his father, in 1807 or 1808, built the first cotton mill in Oneida County, New York, four miles west of Utica.

Some years later Mr. Wolcott, associated with Benjamin and Joseph Marshall, formerly English merchants in New York, built the "New York Mills."

Meanwhile the attention of other parts of the country was being drawn to the subject, and the Society for the Establishment of Useful

Manufactures in New Jersey was organized at New Brunswick, November 22, 1791. In May, 1792, the society selected the falls of the Passaic as the site of their operations, and named their town Paterson after the governor who signed their charter. At a meeting of the directors at the Godwin Hotel July 4th, they made appropriations for building factories, machine-shops, and print-works, and a raceway was directed to be made for bringing the water from above the falls to the proposed mills. Unfortunately the direction of their water-power was given to Major L'Enfant, a French engineer, and the same one who laid out the city of Washington, and his gigantic schemes, reaching from above the falls to tide-water, proved far beyond the means of the company, so that in 1793 the business was put in charge of Peter Colt, then Comptroller of the State of Connecticut, who completed the raceways, abandoning the outlet to tidewater, and built a factory in which they commenced spinning cotton yarn in 1794; and during the years 1795 and '96 much yarn was spun, and several species of cotton fabrics were made. But not succeeding financially, they resolved in July, 1796, to discontinue the manufacture, and discharged the workmen. This result was produced by a variety of causes. Nearly £50,000 had been lost by the failure of parties to certain bills of exchange purchased by the company to buy in England plain cloths for printing; large sums had been wasted by the engineers; and the machinists and manufacturers imported were presumptuous, and ignorant of many branches of the business they engaged to conduct. The cotton mill of the company was subsequently leased to individuals, who continued to spin candle-wicks and coarse yarn until 1807, when it was accidentally burned, and was never rebuilt.

Between 1801 and 1814 several mill-seats were leased to other parties, and in 1814 Mr. Roswell L. Colt purchased at a low price a large proportion of the shares, and reanimated the association, since when the growth of Paterson has been steady, though largely in other directions than that of cotton manufacturing. Still, much cotton machinery and many valuable inventions have been produced there, and we shall probably have occasion to refer to it again in due order. At present Paterson is distinguished as the chief seat of the silk manufacture of the United States, and contains several large and important locomotive and machine works, as well as the different flax mills of the Messrs. Barbour & Brothers.

William Pollard, of Philadelphia, obtained a patent for cotton spinning December 30th, 1791, which was the first water-frame put in motion in Pennsylvania. But whether he obtained his patterns direct from England, or by the way of Pawtucket, is not certain; and it is doubtful if the machinery was capable of successful operation. At any rate, the enterprise failed, while Slater was making his greatest profits, and its want of success is said to have retarded the progress of cotton-spinning in Philadelphia.

In 1808 the Globe Factory, with a capital of $80,000, was established in the "Northern Liberties" of Philadelphia by Dr. Redman Coxe.

The Arkwright machinery was also introduced very early at Copps Creek, Delaware, by Goodfellow, and also at Kirk Mill, near Wilmington.

In 1790 a person who had been engaged in the Beverly Factory was employed to go to Norwich, Conn., to put in operation some cotton machinery which was understood to be similar to that used at Beverly. This machinery is supposed to have been imported by some means from England. In 1794 another mill was built in the west part of New Haven by John R. Livingston and David Dickson, of New York. In 1807 this was converted into a woolen mill, and since into a paper mill. In 1806 General Humphrey built a mill at Derby, Conn., both for cotton and woolen; and the same year a company was formed, consisting of James, Christie, and William Rhodes, brothers, of Pawtucket, Oziel Wilkinson and his five sons, viz., Abraham, Isaac, David, Daniel, and Smith Wilkinson, and his two sons-in-law, Timothy Green and William Wilkinson, with a capital of $60,000, five twelfths of which was invested in real estate then known as Congers Mills, on the Quinebaug River, and included about one thousand acres of lands lying in the adjoining towns of Pomfret, Thompson, and Killingly, in Connecticut.

In this year also (1806) Samuel Slater sent for his younger brother John, who came from England, bringing all the latest improvements in the business, and joined with his brother and his partners in building a new establishment in Smithfield, R. I., now known as the village of Slatersville. In June, 1806, John Slater took charge of the business at this place, under the firm of Almy, Brown & Slaters, and commenced spinning in 1807, managing the business successfully for upward of fifty years. Samuel Slater's business was now becoming very profitable, and he was evidently accumulating property, although his salary as superintendent of the two mills at Pawtucket was only $1.50 per diem from each mill.

The second cotton mill in Massachusetts was built on Bass River, in Beverly, in 1801, with six water-frames of seventy-two spindles each. This machinery was built at Paterson, N. J., by a man named Clark, who came to Beverly to put it in operation. This mill was unsuccessful, from insufficient water-power and other causes, and continued in operation but two or three years.

In 1803 the first factory established at Beverly, having sunk half its capital, suspended operation.

The return of exports for this year shows of Sea Island cotton 8,787,659 lbs., and of other kinds 29,602,428 lbs., and the quantity manufactured in the United States is said to have been 1,000 bales, or 300,000 lbs., as bales then averaged. The prices of yarn at Pawtucket were as follows : No. 12, 99 cents per lb.; No. 16, $1.15 per lb.; No. 20, $1.31 per lb. About this time the first regular cotton factory in the State of New York was erected in Union Village, Washington County, by William Mowry, who had learned the business at Pawtucket.

February 27, 1807, an exemption from taxes for five years was granted by act of the Legislature of Massachusetts for a cotton mill erected at Watertown by Seth Bemis and Jeduthan Faller, and June 20 of the same year a factory was incorporated at Fitchburg, Mass.; and March 12, 1808, the Norfolk Cotton Factory at Dedham was incorporated. A small cotton factory was also established at Pittsburg, Pa., in 1807 ; and Mr. Zachariah Allen estimates the whole number of cotton spindles in the United States to have been about 4,000.

The second cotton mill in New Hampshire was commenced upon the same stream with the first one, the Souhegan River, at New Ipswich, in 1807, and put in operation in 1808 by Seth Nason, Isaac Holton, and Samuel Batchelder, containing, like the first mill, about 500 spindles. In 1805 the Legislature of New Hampshire granted to the proprietors of the first mill an exemption from taxes for five years, and in 1808 the same to the proprietors of the second mill.

I here extract from " The Textile " the following from a letter from Mr. Batchelder to the editor : "Six or seven years before the commencement of weaving by the power loom at Waltham, I was the owner, with two or three others, of the second cotton mill that was built in New Hampshire, and in order to dispose of my part of the product of the mill I undertook to manufacture yarn by the hand loom into shirting, gingham, checks, and ticking. At that time almost every farmhouse in the country was furnished with a loom and spinning wheels, for manufacturing the ordinary clothing of the family, and most of the females were weavers or spinners, and were very willing to undertake to weave such articles as I proposed, in order to purchase calicoes and such other goods as they could not manufacture themselves.

" Before the War of 1812 I made a contract with the other owners of the mill to purchase the whole of the yarn produced by the mill for several years, and extended the business of weaving so that at times I had about a hundred weavers in my employ—not constantly at work, but as they had leisure from other household employment. They came from the neighboring towns for the distance of six or eight miles for the yarn and to return the webs. The price for weaving the different articles was from three to seven cents per yard. On the power loom at the present time the cost would average about one cent. I also at this time made an experiment of weaving on the hand loom pillow-cases without seams, in the manner which was patented many years afterward for weaving bags for grain, which has now become an extensive business. I continued the business for several years after the introduction of the power loom at Waltham, which was confined to weaving plain sheetings and shirtings, while most of the goods which I made were twilled or checks, such as were not woven on the power loom, and consisted in part of dyed yarn of blue and other colors. I paid at that time fifty cents per lb. for dyeing a *fast indigo blue*, such as would now cost only seven or eight cents.

" On looking back at my account books I find that I manufactured more than fifty tons of cloth of various kinds by hand looms, which I continued till 1825, when I went to Lowell to build the Hamilton Mills, where I adopted the power looms for the purpose of weaving twilled goods, such as I had formerly made on the hand loom. My goods were mostly sold in Boston, after commission houses were established for the sale of American goods. Mr. Nathan Appleton states, in his ' Account of the Introduction of the Power Loom,' that on his first bringing the Waltham sheetings to market (1815) there was but one place in Boston where domestic goods were sold; and when, before the War of 1812, I first offered my hand-loom goods for sale in Boston, and proposed to consign them to some dry-goods merchants, I was told that it would be discreditable for them to undertake the sale of American goods, and I had to consign them to retail shops in Salem and other places at a limited price, paying a commission of ten per cent. At one time such was the demand for goods that speculators came from Boston and cleared my shelves of goods at the retail price. Ticking, such as would now be worth fifteen to twenty cents, then sold for seventy-five cents a yard, and a better article sold regularly for a dollar."

In December, 1808, the State of New Hampshire, by a general law, granted exemption from taxation for five years to those who should erect works for the manufacture of cotton, wool, salt, or glass, and incorporated a cotton manufactory at Peterborough, and one at Exeter. In 1809 were incorporated the second Peterborough co. ton factory and one in Chesterfield. In 1810 one was incorporated at Milford, one at Swanzey, one at Pembroke, and one at Amoskeag

Falls, being the nucleus from which has grown the present Amoskeag Company; in 1801, one at Walpole, one at Hillsborough, one at Meredith, and a third at Peterborough. Most of these mills went into operation within about a year of the time of incorporation, so that at the commencement of the War of 1812 there were probably fifteen cotton mills in operation in New Hampshire, averaging not more than five hundred spindles in each, or not more than six or seven thousand in all. The first cotton mill in the State of Maine was built at Brunswick in 1809, and soon after another was erected at Gardiner.

Tench Coxe, in his report of the census of 1810, gives the number of cotton factories as follows:

New Hampshire....... 12	New York............ 26	Maryland............. 11
Massachusetts........ 54	New Jersey........... 4	Ohio................. 2
Rhode Island......... 28	Pennsylvania......... 64	Kentucky............. 15
Connecticut.......... 14	Delaware............. 3	Tennessee............ 4
Vermont............. 1		

This, however, does not agree with other authorities.

Dr. Bishop, in his "History of American Manufactures," gives eighty-seven mills, which he locates as follows:

Maine................ 1	Vermont............. 4	Maryland............. 5
New Hampshire....... 6	New York............ 6	Virginia............. 1
Massachusetts........ 15	New Jersey.......... 2	South Carolina....... 1
Rhode Island........ 25	Pennsylvania........ 4	Georgia.............. 1
Connecticut.......... 6	Delaware............ 2	Ohio................. 1
Kentucky............ 6	Tennessee........... 1	

And of these, fourteen were horse mills.

"These eighty-seven mills were expected to employ a capital of $4,800,000, and use 3,600,000 lbs. of cotton, worth $720,000. They would spin in 1811 2,880,000 lbs. of yarn, worth $3,240,000, and employ 500 men and 3,500 women and children."

Mr. Batchelder says: "All the factories built before the war of 1812 were built after the plan first introduced by Slater, with very little modification. His spinning was what was usually denominated the water-frame, built in separate sections of eight spindles each; but before 1808, when the second mill was built in New Hampshire, the spinning-frame, called the 'throstle,' had been introduced, and was adopted in this mill."

The spread of manufactures, due to the restrictions on the importations of goods, and the consequent advance of prices, was now very rapid, and in 1812 there were said to be nearly forty cotton mills in Rhode Island, with about 30,000 spindles, and about thirty mills in

Massachusetts, within thirty miles of Providence, with about 18,000 spindles, amounting in the whole to 48,000 spindles.

"The war with Great Britain in 1812 raised the price of goods to such extravagant rates, and stimulated the building of cotton factories to such a degree, that a list of the mills in and near Providence, including a number in Massachusetts at the close of the war, makes the number of mills ninety-six, and of spindles 65,264; being an average of 680 spindles to a mill, eighteen of the whole number having less than 300 spindles each, and the largest, that of Almy, Brown & Slater, 5,170 spindles."

This brings us to a new era, that of the power loom, and the introduction of what was known as the "Waltham system," where all the processes of manufacturing cloth were carried on under the same roof and by the same management.

With the growth of the cotton manufacture there had as a necessary consequence sprung up shops for building the machinery, and skillful mechanics had been trained. Calico printed with engraved rollers, and by machinery driven by water-power, was produced by Thorp, Siddall & Co., about six miles from Philadelphia, in October, 1810, and in the same year Alfred Jenks, a pupil and co-laborer with Slater, commenced the manufacture of cotton machinery of every description at Holmesburg, near Philadelphia, afterward removing to Bridesburg, where he engaged extensively in the construction of looms.

The era of weaving by power now demands a new chapter, and will bring a new set of inventors on the stage.

CHAPTER V.

In the following pages I shall have occasion to quote frequently from a pamphlet written by the Hon. Nathan Appleton, of Boston, and printed in 1858 in Lowell by the "Proprietors of the Locks and Canals on Merrimac River," as well as from the authorities previously mentioned, commencing with the following summary of the most reliable facts in the history of the power loom as given by Mr. Batchelder: "The first attempt to weave by machinery was made by M. de Gennes. His loom is described in the 'Philosophical Transactions' in the year 1700. About 1765 a weaving factory driven by water was built by Mr. Garside, of Manchester. It was furnished with 'swivel looms,' probably those invented by M. Vaucanson, and described in the 'Encyclopédie Méthodique.' It was worked for a considerable time, but with no advantage, one man being required for each loom."

The prototype of the present loom, however, was the loom invented by the Rev. Edmond Cartwright, an English clergyman, for which he secured his first patent April, 1785. In his own words, "This being done, I then condescended to see how other people wove, and you will guess my astonishment when I compared their easy modes of operation with mine. Availing myself, however, of what I then saw, I made a loom, in its general principles, nearly as they are now made, but it was not until the year 1787 that I completed my invention, when I took out my last weaving patent, in August of that year."

A weaving factory was built at Doncaster, under Arkwright's license, by some of his friends, but was unsuccessful, and another establishment at Manchester, containing 500 looms, was built by Mr. Grimshaw in 1790, but was destroyed by a mob. The invention, however, overcame all opposition, and at the time of Cartwright's death it was estimated that power looms were performing the labor of 200,000 men.

Another loom was invented by a Mr. Austin, of Glasgow, in 1789, and put in operation in 1798, but with what success is not known.

Patents were obtained for power looms by Robert Miller in 1796, and by Toad, of Bolton, in 1803. Mr. Horrocks, of Stockport, took patents for a power loom in 1803 and 1805, and for further improvements in 1815. This seems to have been the first really successful loom, and has now become of general use, as the crank or "Scotch" loom. Great difficulty was, however, experienced in preparing the yarn so as to produce a warp which could be woven by power; but the desired result was attained by William Radcliffe, of Stockport, and Thomas Johnson, of Bredbury, who in 1804 patented the "dressing-machine," and also took out patents for improvements in the loom, taking up the cloth as woven by the motion of the lathe. Horrocks and Radcliffe, like nearly all inventors, failed financially, which retarded the adoption of their inventions, so that in 1813 it was supposed that not more than 100 dressing-machines and 2,400 looms were in operation in Great Britain. Still these were enough to alarm the hand-loom weavers, who, attributing to machinery the distress arising from the American war and the "Orders in Council," broke all the looms set up at Middleton, West Houghton, and other places. (Baines's "History of the Cotton Manufacture," etc.)

Mr. Batchelder says that a loom was built at Exeter, N. H., by T. M. Murphy in 1806, and experimented with for three years, and experiments were also made at Dorchester and Dedham, Mass., between 1806 and 1809, but were not sufficiently successful to take the place of hand-weaving.

"In the year 1811," says Mr. Appleton, "I met my friend, Mr. Francis C. Lowell, at Edinburgh, where he had been passing some time with his family. We had frequent conversations on the subject of the cotton manufacture, and he informed me that he had determined before his return to America to visit Manchester for the purpose of obtaining all possible information on the subject, with a view to the introduction of the improved manufacture in the United States. I urged him to do so, and promised him my coöperation. He returned home, and in 1813 came to me on the Boston Exchange one day, with Mr. Patrick T. Jackson, and stated that they had purchased a water-power in Waltham (Bemis's paper-mill), and that they had obtained an act of incorporation, and Mr. Jackson had agreed to give up all other business, and take the management of the concern." Mr. Jackson was the brother-in-law of Mr. Lowell, and from a memoir of him published in 1858 in Hunt's "Merchant's Magazine," by John A. Lowell, Esq., of Boston, I extract the following:

"Mr. Lowell had just returned to this country in 1812, after a long visit to England and Scotland. While abroad he had conceived the idea that the cotton manufacture, then almost monopolized by Great Britain, might be advantageously prosecuted here. The use of machinery was daily superseding the former manual operations, and it was known that power looms had recently been introduced, though the mode of constructing them had been kept secret. The cheapness of labor and abundance of capital were advantages in favor of the English manufacturers; they had skill and reputation. On the other hand, they were burdened with the taxes of a prolonged war. We could obtain the raw material cheaper, and had a great superiority in the abundant water-power, then unemployed, in every part of New England." "So confident was he in his calculations that he thought he could in no way so effectually assist the fortunes of his relative, Mr. Jackson, as by offering him a share in the enterprise." Mr. Jackson had been engaged in the Calcutta trade, which was prostrated by the war, and gladly took hold of the new enterprise.

The difficulties were very great. The war precluded all communication with England. Neither books, designs, nor models could be procured; everything had, as it were, to be reinvented; and the power loom was the first thing to be accomplished. "In England it had been invented by a clergyman: why not here by a merchant?"

After numerous trials, they succeeded in the autumn of 1812 in producing a model which was so satisfactory that they engaged the services of Mr. Paul Moody, of Amesbury, a well-known and skillful mechanic, to aid in the construction of a mill for weaving cotton cloth. The first project was for a weaving-mill exclusively, but it was found that it would be more economical to spin the yarn than to buy it, and they put up a mill with about 1,700 spindles, which was completed late in 1813. This was probably the first mill in the world that combined all the operations for converting the raw cotton into finished cloth.

Great difficulty was at first experienced at Waltham for the want of a proper machine for preparing the warps. They procured from England a drawing of Horrocks's dressing machine, which they altered and very much improved, producing the "dresser," which has till recently been of general use in this country. No plan, however, was shown in this drawing for putting the yarn on the section beams, and to supply this deficiency Mr. Moody invented the warper.

The " stop motion," to indicate the breaking of any of the threads on the warper, was also invented at Waltham by Jacob Perkins, the inventor of the present system of bank-note engraving on steel, and other ingenious inventions. Other great improvements were made. Mr. Appleton says: "The greatest improvement was in the 'double-speeder.' The original fly frame or roving frame introduced from England was without any fixed principle for regulating the changing movements necessary in the process of filling a spool. Mr. Lowell undertook to make the numerous mathematical calculations necessary to give accuracy to these complicated movements, which occupied him constantly for more than a week. Mr. Moody carried them into effect by constructing the machinery in conformity. Several trials at law were afterward had about this patent, involving, among other questions, one whether a mathematical calculation could be the subject of a patent." The last great improvements consisted in giving a more slack twist to the yarn spun for filling on the throstle, and in spinning it directly on the shuttle "quill" without the process of winding. Mr. Lowell and Mr. Moody went to Taunton to purchase a filling-winder, the patent of which was owned by Mr. Shepherd, of that place, having previously tried one made by Mr. Stowell, of Worcester. Mr. Lowell tried to get a reduction in price from Mr. Shepherd, which he refused, telling them that "they must have the machines, as they could not do without them," when Mr. Moody remarked that "he was just thinking that he could spin the filling direct upon the bobbin." Mr. Lowell, who perceived the practicability of doing this, dropped the subject, and after some further conversation they took leave. On their return, Mr. Lowell told Mr. Moody that he must accomplish the plan he had suggested, and the invention of the "filling frame" was the result. This has been of late years superseded to a great extent by the mule; but, since the recent improvements in ring-spinning, the motions have been applied to the ring-frame, and it has been proved that the coarser numbers of yarn, say below No. 20, can be spun cheaper in this way than on the mule.

A similar jesting remark also led to the adoption of soapstone for the rollers in the dressing-frame, instead of wood, which swelled and warped so much with the moisture of the size that they would not work properly.

Mr. Moody's brother suggested to him the use of a "soapstone" mold in which to cast some "pewter" rollers. Mr. Moody took the hint in a manner different from what was intended, and made the rollers themselves of soapstone.

Mr. Lowell's loom was different in several particulars from the English loom, the principal one being that the lathe was driven by an eccentric cam instead of a crank, which has now been generally substituted for the cam motion, and some other improvements have been introduced.

With the success of the new machinery, there was no difficulty in raising the capital of $400,000 required to carry out the scheme at Waltham. A charter, under the name of the Boston Manufacturing Company, was obtained, and the full water-power utilized, and $200,000 additional were afterward raised for the purchase of the adjoining property in Watertown.

With the mechanical success of Waltham, and the adoption of the new system, which long went by its name, came other great changes, which were equally the result of the foresight and sagacity as well as the philanthropy of Mr. Lowell; and I can not better express them than by the following quotation from the memoir of Mr. Jackson, above referred to.

"It is not surprising that Mr. Lowell should have felt great satisfaction at the result of his labors. In the establishment of the cotton manufacture in its present form, he and his early colleagues have done a service not only to New England, but to the whole country, which perhaps will never be fully appreciated: not by the successful establishment of this branch of industry—that would sooner or later have been accomplished; not by any of the present material results which have flowed from it, great as they unquestionably are; but by the introduction of a system which has rendered our present manufacturing population the wonder of the world. Elsewhere vice and poverty have followed in the train of manufactures; an indissoluble bond of union seemed to exist between them. Philanthropists have prophesied the like result here, and demagogues have reëchoed the prediction. These wise and patriotic men, the founders of Waltham, foresaw and guarded against the evil. By the erection of boarding-houses at the expense and under the control of the factory, putting at the head matrons of tried character, and allowing no boarders to be received except the female operatives of the mill; by stringent regulations for the government of these houses—by all these precautions they gained the confidence of the rural population, who were now no longer afraid to trust their daughters in a maunfacturing town. A supply was thus obtained of respectable girls; and these, from pride of character as well as principle, have taken especial care to exclude all others. It was soon found that an apprenticeship in a factory entailed no degradation of character, and was no impediment to a reputable connection in marriage.

"A factory-girl was no longer condemned to pursue that vocation for life, and it soon came to be considered that a few years in a mill were an honorable mode of securing a dower. The business could thus be conducted without any permanent manufacturing population. The operatives no longer form a separate caste, pursuing a sedentary employment, from father to child, in the heated rooms of a factory, but are recruited in a circulating current from the healthy and virtuous population of the country. By these means, and a careful selection of men of principle and purity of life as agents and overseers, a great moral good has been obtained. Another result has followed, which, if foreseen, as no doubt it was, does great credit to the sagacity of these remarkable men. The class of operatives employed in our mills has proved to be as superior in intelligence and efficiency to the population elsewhere employed in manufactures as they are in morals. They are selected from a more educated class, from among persons in more easy circumstances, where the mental and physical powers have met with fuller development. This connection between morals and intellectual efficiency has never been sufficiently studied. The result is certain, and may be destined in its consequences to decide the question of our rivalry with England in the manufacture of cotton."

I have quoted thus at length from this memoir, written nearly thirty years ago, because the ideas expressed in it seem to me to be yet worthy of careful study; although the special manufacturing population, which the founders of Waltham so much deprecated, is growing up among us, under the influence of competition, combined with an enormous foreign immigration, and the growth of our manufacturing towns, which have increased so as to form the homes of a permanent population. Under the circumstances, the policy pursued by Mr. Lowell and his associates was not only wise, but necessary. The water-powers which it was proposed to use in developing the new industry on a grand scale were literally "in the woods." Dwellings for the operatives were to be constructed in these solitudes, and the operatives to be procured. The latter were readily found in the surplus female population scattered all over New England, many of whom had learned to spin and weave on the hand-wheel or loom; and philanthropy and economy went hand in hand in the organization of the new system. The church and factory were built together, and the school-house soon followed.

From Waltham this mode of organization spread rapidly to the northern district of New England—Lowell, Saco, Dover, etc.—which will be noticed in due course; but chronological order compels us to return to Rhode Island and its vicinity; and in this connection it is but proper to say that Slater and his associates also established Sunday-schools, and took great interest in the moral welfare of their employees, though the permanent or family system was adopted by them in the small way in which their mills were commenced as compared with the scale of operations at Waltham and afterward at Lowell. There were also other great points of difference between what we may call the "Waltham" and the "Rhode Island" systems. At Waltham the wages were paid in cash, no children were employed, and the operatives were free to make their purchases at their discretion. In Rhode Island the owners of the mills established a sort of "factory store" from which the families were supplied with all that they needed on credit, and but little cash was used in the daily transactions, and the employees were kept in a sort of dependence on their employers.

CHAPTER VI

In 1812 the first cotton mill in Fall River, then called Troy, was erected at what was known as Globe Village by Colonel Joseph Durfee and others, and was afterward converted into the Globe Print Works in 1829, and was burned in December, 1838. In 1813 the Troy Manufacturing Company was organized and built their first mill. From that time until 1840 the growth of the place was slow, and in the latter year there were 32,864 spindles where there are now over 1,200,000. I merely note its commencement here in due order. In 1814 manufacturing was again revived at Paterson, as noted by Mr. Roswell L. Colt, the son of Mr. Peter Colt; and it was estimated that Essex County, N. J., contained in September of this year 32,500 spindles. The second steam-engine in Providence, of 24 horse-power, was erected in this year by Oliver Evans, for the Providence Dyeing, Bleaching, and Calendering Company.

The Bellingham Cotton and Woolen Factory, on Charles River, and the Hampden Cotton Manufacturing Company, on Chicopee River, at Ludlow, were incorporated by Massachusetts. The first cotton mill in North Adams, Mass., was erected in 1811 and started in 1812. The cotton was received here as it was picked in the cotton-field. It was put out in families and picked and whipped, then sent back to the mill and carded and spun into yarn. This yarn was put out in the families by the company, and woven on hand looms into blue and white striped cloth for pants, also another style for women's dresses. The name of this company was "The Adams North Village Cotton Factory Company." The Eagle Mill was started in 1814, near the Eagle Mill of the Freeman Manufacturing Company, and in 1819 the first power loom was started in that mill. The first power loom on satinet was started by S. Burlingame in 1823. The first cotton mill at Fishkill, N. Y., the commencement of the Mattcawan Manufacturing Company, was built by Messrs. Schenck & Dowling, and the Lancaster Manufacturing Company, of Lancaster, Penn., commenced operations, but failed in 1818.

In 1815 William Gilmore came from Scotland to this country, arriving in Boston in September. He was met in Boston by Mr. Robert Rogerson, who knew that he had been employed in power-loom weaving, and understood the construction of the looms and dressing machinery, and who took him to Smithfield and introduced him to John Slater. He proposed to Mr. Slater to build the machinery for power-loom weaving, receiving nothing for his labor unless he succeeded in putting the looms in operation. But the prospects of business were at that time so discouraging that parties were not willing to enter into engagements, and he went to work as a machinist at Smithfield, where he commenced paying rent October 21, 1815. Previous to this time a machinist by the name of Blydenburg had been employed at the Lyman Mills, in North Providence, in the attempt to build a power loom, but so far without success. Gilmore was employed, in the early part of 1816, to build twelve looms, and also machinery for warping and dressing, from the plans and drawings which he had brought with him, which he accomplished to the satisfaction of his employer, and they were put in operation early in 1817.

For the compensation of ten dollars he allowed Messrs. David Wilkinson & Co. the use of his patterns for building twelve other looms; and they got their looms in operation nearly as soon as those built by Gilmore. This was the first introduction of the crank loom in this country; and, to manifest their gratitude for the services rendered by Mr. Gilmore, the manufacturers subscribed to raise a fund of $1,500, and one of the subscribers to this fund refers to his receipt for payment of his subscription, which he has preserved, dated May 31, 1817, thus showing the time when the crank loom was put in operation in this country.

Mule spinning having been introduced in Rhode Island, the building of the power loom, as noted, completed the manufacturing system of that State within about three years after the power loom was built at Waltham.

In order to avoid the use of the patented machinery used at Waltham, the Rhode Island mills adopted the crank loom, and introduced various roving frames copied from English models, among them, at a later date, the tube speeder, invented by George Danforth, of Massachusetts, and otherwise known as the "Taunton speeder." This was also introduced to a considerable extent in Great Britain in 1825. Another form of roving frame was known as the "Brown" speeder, patented in 1821 by John Brown, of Providence. But the two systems differed essentially in the fact that the Rhode Island district adopted the "live" spindle, the Waltham and Northern district the "dead" one; the first, the mule, Scotch dresser, and crank loom; the other, the filling frame, Waltham dresser, and cam loom.

An important addition to the loom was made in 1816 by the invention of the "rotary temple" by Ira Draper, of Weston, which was introduced in this country many years before it was adopted in England, where they clung for a long time to the use of the old hand "temple," for keeping the cloth extended after leaving the reed. This temple was afterward improved by his son, George Draper, who carries on the only manufactory of temples in the United States, and to whom we shall again have occasion to refer in connection with other valuable inventions.

Cotton machinery as perfect as any in existence was now fairly introduced into America, and during the war the number of spindles in New England had increased to over 120,000. A statement compiled by Samuel Green, of Woonsocket, which I copy from Mr. Batchelder, and which was made for "The Rhode Island Society for the Encouragement of Domestic Industry," gives the statistics as follows in 1815:

```
Rhode Island..................................  99 mills, 68,142 spindles.
Massachusetts.................................  52   "    34,468     "
Connecticut...................................  14   "    11,700     "
                                                ---       -------
                                                165       119,310
```

This says nothing about Maine or New Hampshire.

I also copy the following table from a report of the Committee on Manufactures to Congress in 1815:

```
Capital...............................................................  $40,000,000
Males employed, of the age of seventeen and upward............            10,000
Boys under seventeen .........................................            25,000
Women and female children.....................................            66,000
Wages of 100,000, average $1.50 per week............  ........           $7,800,000
Cotton manufactured, 90, 00 bales, or lbs..................=             27,000,000
No. yards cloth.......................................................   81,000,000
Cost, averaging 30 cents per yard..........................              $24,300,000
```

The close of the war in 1815, and the consequent enormous increase of importations, threatened ruin to this newly born and imperfectly developed interest, and in 1816, after a long struggle, the tariff act of that year was passed, recognizing the principle of "protection to American industry." A duty of 25 per cent. *ad valorem* was levied on all cotton cloths, and the minimum valuation at the port of exportation was fixed at 25 cents per square yard, thus amounting to a specific duty of 6¼ cents per square yard. This rate was to be leviable for three years, after which it was to be reduced to 20 per cent. *ad valorem*, and the same rates were to be applied to cotton twist, yarn, or thread, unbleached costing less than 60 cents per lb., and bleached or colored less than 75 cents per lb.

With the help of this protection by Government, and the introduction at the same time of the power loom, the cotton manufacture became established as a national industry, and has continued to thrive, though subject to great fluctuations, from the uncertain and varying policy of Congress in regard to the tariff on imported goods.

Mr. Appleton says : "By degrees the manufacturers woke up to the fact that the power loom was an instrument which changed the whole character of the manufacture, and that, by adopting the other improvements which had been made in machinery, the tariff of 1816 was sufficiently protective."

The Legislature of New Jersey in 1815 abolished the tax on spindles employed in the cotton manufacture. Twelve hundred spindles are said to have been in operation in Cincinnati this year. There was one manufactory of fustians and cotton velvets at Hudson, N. Y., and one about to be started at Frankfort, Penn.

In 1816, Mr. Seth Bemis, of Watertown, Mass., applied the power loom to the manufacture of cotton duck, which he had commenced with the hand loom in 1809, and which was made from Sea Island cotton, costing then from 20 to 25 cents per lb. ; and in this year also Jephtha Wilkinson, of Otsego, N. Y., patented a machine for making loom reeds. In 1817 "societies for promoting American manufactures" were formed in a number of the States. In 1818 the first cotton factory in North Carolina was established at the Falls of Far River, in Edgecombe County, and was followed by another near Lincolnton, on the Catawba, in 1822. "In 1819 the first cotton mill in Manayunk, Philadelphia, was built by Captain John Towers, and afterward passed into the hands of Joseph Ripka, to whose enterprise the growth of that place is principally due." (Bishop, vol. ii., p. 253.)

During these years, 1817–'20, the cotton manufacture had, however, been in a declining condition, and urgent and incessant endeavors had been made to secure further legislation by Congress, but without success. The report of a committee of that body, based on the census of 1820, shows the pounds of cotton actually spun in that year to have been 9,945,609, being a decrease of 63 per cent. on the amount consumed in 1815, and distributed as follows :

	No. Spindles.	Lbs. Cotton spun.
Maine	3,070	56,500
New Hampshire	13,012	412,100
Massachusetts	80,304	1,611,796
Rhode Island	63,372	1,914,220
Connecticut	29,826	897,335
Vermont	3,278	117,250
New York	33,160	1,412,495
New Jersey	18,124	648,600
Pennsylvania	13,776	1,062,753
Delaware	11,784	423,800
Maryland	20,245	849,000
Virginia		3,000
North Carolina	288	18,000
South Carolina	588	46,449
Kentucky	8,097	360,951
Ohio	1,680	81,360
Total	250,572	9,945,609

In 1821 the cotton crop of the United States had increased to 180000,000 lbs., of which 124,000,000 lbs. were exported. Secretary Woodbury estimated the amount consumed in the United States to have been 20,000,000 lbs., which, compared with the census returns just quoted, leaves about 10,000,000 lbs. to have been manufactured by hand labor.

Despite this general prostration of the manufacturing interest, arising from a variety of causes, into which I have neither space nor time to inquire, the factory at Waltham, which was the largest one in the country, had been uniformly successful, whether from its size and ample capital, or its more perfect organization and the great business talent of its owners, is not for me to say; and is said to have paid twelve per cent. annually during this period of depression, and its principal owners, Messrs. Jackson & Appleton, began inquiries in 1821 for a water privilege where they could commence the manufacture and printing of calicoes on a large scale. And to the result of their search and the foundation of Lowell I will devote another chapter, having in this given all the statistics which I can procure of the progress of the manufacture up to this time.

CHAPTER VII.

The year 1821 is a memorable one in the annals of American manufactures as witnessing the inception of a larger enterprise than had yet been attempted, and which, in view of its full results, may be fairly termed "gigantic"—the foundation of the city of Lowell. Mr. Nathan Appleton says: "I was of opinion that the time had arrived when the manufacture and printing of calicoes might be successfully introduced into this country. In this opinion Mr. Jackson coincided, and we set about discovering a suitable water-power. At the suggestion of Mr. Charles H. Atherton, of Amherst, N. H., we met him at a fall of the Souhegan River, a few miles from its entrance into the Merrimac; but the power was insufficient for our purpose. This was in September, 1821. In returning we passed the Nashua River, without being aware of the existence of the fall which has since been made the source of so much power by the Nashua Company. Soon after our return, I was at Waltham one day, when I was informed that Mr. Moody had lately been at Salisbury, when Mr. Ezra Worthen, his former partner, said to him, 'I hear that Messrs. Jackson & Appleton are looking out for water-power. Why don't they buy up the Pawtucket Canal? That would give them the whole power of the Merrimac, with a fall of over thirty feet.' On the strength of this, Mr. Moody had returned to Waltham by that route, and was satisfied of the extent of the power which might thus be obtained, and Mr. Jackson was making inquiries on the subject."

Mr. Jackson a day or two after called on Mr. Appleton and explained his plans, which were carried out by associating with themselves Mr. Kirk Boott, a merchant of Boston who had been educated in England, and had some knowledge of engineering, and who was desirous of engaging in the active management of the enterprise. Mr. Thomas M. Clark, of Newburyport (the father of the present Bishop of Rhode Island, and the agent of the Pawtucket Canal Company), was employed by these gentlemen jointly to buy up all the lands about the falls and canal, and such shares of the Canal Company as were within his reach; while Mr. Henry M. Andrews was employed to purchase all the shares owned in Boston. All this was done very quietly in order to prevent attempts at extortion or speculation; and in November, 1821, Messrs. Jackson, Appleton, Boott, Dutton, and Moody visited the spot. Formal articles of association were then drawn up, bearing date December 1, 1821, and in 1822 the Merrimac Manufacturing Company was incorporated, on the 5th of February. On the 27th of February the first meeting of stockholders took place, and a board of directors was chosen, to whom was transferred the property which had been purchased, and for which had been paid, viz., for lands, $18,339, and for 339 canal shares, $30,217. The Pawtucket canal had been originally built to facilitate the navigation of the Merrimac River, and its enlargement and the renewal of the locks was the first thing to be done. This was commenced in 1822, and completed in 1823 at a cost of $120,000, and was estimated to furnish fifty "mill-powers." This term of "mill-power," or "mill-privilege," used in Lowell, Lawrence, and all the northeastern district, is derived as follows: "The second mill built at Waltham contained 3,584 spindles, with all the apparatus necessary to spin No. 14 yarn, and convert it into cloth, which was taken as a standard, and the necessary water-power was estimated and established as the right to draw twenty-five cubic feet per second on a fall of thirty feet, or a gross horse-power of 85.05, supposed to net about 60 horse-power. The price for this was fixed at Lowell at $4 per spindle, or $14,336 for a mill-power and the necessary land, of which $5,000 were to remain unpaid, subject to an annual rent of $300, or $5 per horse-power." This quantity of water, or its equivalent, according to the height of the fall used, has been ever since adopted as the standard in the towns which have followed Lowell, and the water rent has also been substantially the same. An arrangement was made with the Waltham company to equalize the interest of the stockholders in both companies, by mutual transfers at rates agreed on, and to pay the Waltham company $75,000 for all their patterns and pattern-rights, and the release of Mr. Moody from their service, as he was indispensable to the new company.

Houses for Mr. Boott, Mr. Moody, and the operatives were built, as well as the first Merrimac Mill, and a church, and the first water-wheel was started September 1, 1823. Mr. Worthen was the first superintendent of the mills, and Mr. Moody moved from Waltham, and took charge of the machine-shop. Print-works were also commenced in 1823, and at the same time were begun by the Dover Manufacturing Company at Dover, N. H., and also at Taunton, Mass. The Great Falls Manufacturing Company, at Somersworth, N. H., was also incorporated this year, as well as the Newmarket Manufacturing Company at Newmarket, N. H. Mr. Worthen, who was a man of great mechanical skill and ingenuity, died very suddenly in 1824, deeply regretted by all with whom he had been connected, and was succeeded by Warren Colburn, the mathematician.

"The original capital of $600,000 was increased to $1,200,000 in 1823, and in October, 1824, a new subscription of 600 shares was voted, and a committee appointed to consider the expediency of organizing the Canal company by selling them all the land and water-power not required by the Merrimac Manufacturing Company. This committee reported on the 28th of February, 1825, in favor of the measure, which was adopted, and at the same time a subscription was opened by which 1,200 shares in the locks and canals were allotted to the holders of that number of shares in the Merrimac Company, share for share."

In 1821 Mr. Paul Moody had taken out patents for improvements in spinning-frames, and two patents for roving-frames ; one of them being the "double-speeder." These and other improvements were introduced into the new factories with great advantage. The business of printing calicoes was entirely new in this country, and required some time to establish in a satisfactory manner. The print works of the Merrimac Company were at first placed under the charge of Mr. Allen Pollock, but in 1826 Mr. John D. Prince, of Manchester, England, was induced to come out to this country and take the charge of them, bringing with him from England the knowledge of the art of printing by engraved copper cylinders, by machinery, then just introduced there by Robert Peel, and which was taking the place of the old method of printing by hand with wooden blocks. Dr. Samuel L. Dana, of Waltham, the distinguished chemist, was also engaged by the company ; and by the joint skill and talent of Messrs. Boott, Prince, and Dana the success and reputation of the Merrimac Company were established.

The first mills built by them were about 150 feet long by 45 wide, and five stories high, containing about 6,000 spindles each, with the necessary preparation and looms, and of these there were five, making about 30,000 spindles, which was their full complement for some twenty years. Two of the original mills have been destroyed by fire, and the others torn down and rebuilt on a larger scale ; so that the five mills now standing contain, by the Lowell statistics of 1876, 158,-464 spindles and 3,941 looms, with a capital of $2,500,000, and employ 900 male and 1,800 female operatives.

The Locks and Canals company now organized, taking the machine-shop as part of their property, and built the necessary new canals to improve the remaining water-power, and in 1825 made their first sale to the Hamilton Manufacturing Company, who secured the services of

Mr. Samuel Batchelder, of New Ipswich, N. H., who had shown great manufacturing skill, as their first agent, and under his management the power loom was applied with great success to the weaving of twilled and fancy goods. In 1828 the Hamilton Company commenced calico printing, under the charge of Mr. William Spencer, who came out here from England for that purpose.

In 1828 the Appleton Company was organized, and their mills contained various later improvements by Mr. Moody, and are believed to have been the first in which the system of driving the whole mill by "main belts" instead of geared shafting was put in operation, and which was the suggestion of Mr. Moody, who thereby saved, not only in first cost, but in power.

All these early mills in Lowell were of about the same size, i. e., 5,000 to 6,000 spindles, and I will now only enumerate the order in which the remaining companies were organized, giving their present statistics in their proper order in my summary. The Lowell Manufacturing Company, for coarse cottons, negro cloths, and carpets, was incorporated in 1828; the Suffolk and Tremont Mills, now united, in 1830, commencing the manufacture of heavy drills, which experiments by Mr. Batchelder at the Hamilton Company had shown to be very profitable. With these mills commenced the connection of Messrs. Amos and Abbott Lawrence with the manufacturing interest, and in 1831 the Lawrence Manufacturing Company was incorporated, commencing operations in 1833. The Boott Cotton Mills, on a somewhat larger scale, followed in 1835, and the Massachusetts Cotton Mills in 1839; succeeded by the Prescott Mills (now united with the Massachusetts) in 1844.

The Lowell Bleachery was incorporated in 1832, and the Lowell Machine-Shop was also incorporated as a separate establishment in 1845, when the Locks and Canals Company sold out the remainder of their real estate, and was reorganized as a water-power company only, under the charge of James B. Francis, Esq., the eminent hydraulic engineer, who had been for a long while the engineer of the old company. The proceeds of the sales of the shop and real estate were divided among the old stockholders, and the new stock was taken *pro rata* by the different manufacturing companies, working under the original charter of 1792, as "the Proprietors of the Locks and Canals on Merrimac River."

The Middlesex Company for the Manufacture of Woolen Goods, incorporated 1830, completes the list of the larger corporations in Lowell.

The old canals proving insufficient for the proper supply of water to the mills, a new and much larger one was constructed by Mr. Francis, in 1846, by the erection of a massive stone wall for a long distance parallel with the old bank of the river, and founded on its rocky bed; then turning, was excavated through the ledge, and carried into the heart of the city. The cost of this work was about $500,000. During the past year a massive and permanent stone dam, laid in cement, has been built by Mr. Francis just below the original one, and the water-power of Lowell may be considered as completed. In March, 1826, the *town* of Lowell was set off from Chelmsford, and in 1836 it was incorporated as a city.

Meanwhile, there had been growth elsewhere. In 1822 the manufacture of cotton sail-duck was commenced in Paterson, N. J., by John Colt, with hand looms, for which Mr. Bemis's power loom was substituted in 1824, and the business was rapidly extended, and was established in Baltimore in 1823 by Charles Crook, Jr., & Brother, who were themselves unsuccessful in business, but the manufacture of duck has remained and become prosperous in Baltimore. In 1822 the first "Texas" cotton appeared in market. In 1823 the Nashua Manufacturing Company was incorporated in New Hampshire, and the Dover Manufacturing Company, at Dover, incorporated in 1822, which failed, and was merged in the Cocheco Manufacturing Company in 1827, had commenced operations, besides mills at Exeter and Pembroke, N. H.; while in Massachusetts the Boston and Springfield Manufacturing Company, with a capital of $500,000, was chartered, and commenced operations on the Chicopee River, in the town of Springfield. The name of this company was changed in 1828 to that Chicopee Manufacturing Company.

It is now impossible to trace the exact progress in different localities—it was so rapid and various. The Blackstone Manufacturing Company in Massachusetts commenced about this time, as well as the Coventry Manufacturing Company in Rhode Island, according to Bishop (vol. ii., p. 284).

In 1824, after a long and tedious struggle, a new tariff bill was passed by Congress, which raised the minimum valuation on cotton cloths from twenty-five to thirty-five cents the square yard, and also levied a duty of three and three quarters cents per square yard on cotton bagging, for the benefit of Kentucky and the Western States. A report of the Secretary of State, in answer to a resolution of the Senate, this year, gives the manufacturing capital authorized by State laws since 1820 as, in New Hampshire, $5,830,000; in Massachusetts, $6,840,000; in Connecticut, $1,300,000; and New York, $797,000—making a total in seven States, with the amount authorized in 1820, of $70,636,500. This, however, includes other industries besides cotton. A bleaching and print works was incorporated at Belleville, N. J., this year, and Philadelphia was estimated to contain thirty cotton-mills, averaging 1,400 spindles each. The Ware Manufacturing Company was also incorporated at Ware, Mass.

Numbers of patents for improvements in cotton machinery had also been issued since the establishment of power-loom weaving at Waltham, one of the most important of which was that of the application of the "compound gear," "differential box," or "equation box" to the roving-frame. This motion, which has been adopted universally in all roving-frames or speeders, and which was patented in England in 1826, by Henry Houldsworth, Jr., was the invention, according to Mr. Zachariah Allen, of Providence, of Aza Arnold, a native of Rhode Island, who first applied it in 1822, but took no measures to secure his patent till January 7, 1823.* A model was taken to England by an American in 1825, and it was seized upon eagerly and came into general use; English writers giving the credit of the invention to Mr. Houldsworth. This beautiful invention for regulating the different velocities of the spindle and bobbin, so that the surface of the bobbin, while constantly increasing in size, still preserves the same relation to the speed of the rollers, consisted of a train of bevel gears, one of which was set in the web of another gear, and, while possessing a rotary motion of its own, also revolved bodily around another center; thus either retarding or accelerating the motion transmitted through it, according to the direction given to the intermediate gear. A simple change of a single pinion, which gave this regulating motion, at once adapted the frame to any size of roving, without the great cost which attended the alteration of the old speeder to different numbers from those for which it was originally geared. Mr. Arnold's neglect in asserting his patent led to tedious and expensive litigation; so that he reaped no proper reward for his ingenuity, while the parties who appropriated it in England are said to have derived great profit.

The year 1825 is marked by the invention by Richard Roberts, of Manchester, England, of the self-acting mule, which was not, however, introduced into the United States until some years later.

The amount of cotton grown this year in the United States is stated at 255,000,000 lbs., of which 176,500,000 lbs. were exported, and the number of spindles in the country is given as 800,000. The first commencement of operations at Saco, Me., was made this year by a company of Boston gentlemen who bought Cutts's Island, at the mouth of the Saco River, and formed their plans for a cotton-mill of 12,000 spindles, which was built in 1826, and was the largest mill yet attempted in America.

* See Appendix B.

Bishop (vol. ii., pp. 308 and 309) gives an estimate for 1826 of 400 cotton factories in New England, averaging 700 spindles each, and consuming 98,000 bales of cotton. These were distributed as follows: Massachusetts, 135; Rhode Island, 110; Connecticut, 80; New Hampshire, 50; Maine, 15; Vermont, 10. The larger villages in the order of their size he gives as follows: Lowell, Mass.; Great Falls, Dover, and Nashua, N. H.; Pawtucket, R. I.; Fall River, Mass.; Blackstone, Mass.; Slatersville, R. I.; Taunton, Mass.; Pawtuxet, R. I. (?); Ware and Waltham, Mass.; New Ipswich and Newmarket, N. H.; Springfield and Lancaster, Mass.; Norwich, Conn. The number of cotton factories in the other States was estimated at 275, of the same average size; making the total consumption of cotton 150,000 bales per annum.

This year the Hudson Calico Print-Works were established at Columbiaville, near Hudson, N. Y., by Joseph and Benjamin Marshall, and have since been largely increased; and the Cohoes Company was incorporated, for the improvement of the water-power of the Mohawk River where it falls into the Hudson. For the following statistics I am indebted to D. J. Johnston, Esq., of the Harmony Mills. The first attempt at manufacturing at Cohoes was made in 1811, when the Cohoes Manufacturing Company was incorporated for the purpose of manufacturing cotton, woolen, and linen goods. This came to nothing; and in 1826 the Cohoes Company, with a capital of $250,000, since increased to $500,000, was formed, as noted. They built a dam and canals, making the whole fall of 103 feet available on five different levels. The Harmony Manufacturing Company was incorporated in 1835 by a company of gentlemen from Albany and New York, among whom was Peter Harmony, from whom the company received its name. They erected in 1837 Mill No. 1, which still stands, enlarged and improved. Financially this concern proved unsuccessful, and in 1850 it passed by a compulsory sale into the hands of the present company; the present statistics of which show as follows: Mill No. 1, 550 by 70 feet, 4 stories, containing 42,000 spindles and 900 looms; No. 2, built in 1857, enlarged in 1866, is 660 by 75 feet and 3 stories, containing 48,500 spindles and 1,100 looms; No. 3, built 1866, enlarged 1872, 1,250 by 76 feet, 5 stories, containing 130,000 spindles, 1,700 narrow and 1,100 wide looms; No. 4, formerly the "Ogden Mill," is 500 by 50 feet, 5 stories, containing 29,000 spindles, 650 looms; No. 5, formerly the "Strong Mill," 330 by 50 feet, 4 stories, containing 19,000 spindles, 330 looms; No. 6, 240 by 60 feet, 3 stories high, containing 4,000 spindles and 100 looms. This latter mill is used for the manufacture of jute goods, seamless bags, and coarse yarns. The aggregate amount of machinery used by this company is

267,500 spindles and 5,880 looms; more than the whole estimate for the United States in 1820.

For the first time in the returns of exports, manufactured cotton goods are included in 1826, amounting to a value of $1,138,125, of which $671,266 was sent to Mexico and Central and South America.

The manufacture of cotton bagging was also attempted this year (1826), at Nashville, Tenn. In 1827 the Cocheco Manufacturing Company bought for $750,000 the property of the Dover Company, which had cost about $2,000,000, and continued operations at Dover.

The cotton manufactures of Philadelphia and vicinity for this year are stated by Dr. Bishop to have amounted to 24,300,000 yards, worth $3,888,000, and consuming 20,250 bales of cotton, and the total consumption of cotton for the United States is estimated at 103,482 bales. Six thousand bales were consumed this year at Paterson, N. J. In 1827 or 1828 subscriptions were made for the first cotton factory in Virginia, at Petersburg, and a company was also projected about this time at Fredericksburg.

The year 1828 is memorable for the passage of a positively "protective tariff" by Congress, which, however, especially favored the woolen and iron interests.

On the 2d of September, 1828, Charles Danforth, of Paterson, N. J., received his patent for the "cap spindle," which was introduced into England in 1830, and extensively used, both there and at home, for spinning "filling" before the great improvements in and general introduction of the self-acting mule, and is now being adopted successfully for spinning worsted "filling" in this country.

CHAPTER VIII.

ANOTHER patent was granted this year (1828) to John Thorpe, of Providence, which has been productive of such enormous results, and so changed the character of cotton spinning in America, that it deserves to head a chapter. I refer to the "ring spindle," which has driven out the "cap," and almost superseded the throstle in the manufacture of cotton warp-yarns.

Dispensing with the flier, which carried the yarn around the bobbin in the original patent, the yarn was led in by a slot in the outer one, between two concentric rings, around the inner one of which it was carried by the revolutions of the spindle, one or two modifications of which were shown in the drawings.

In 1829 a patent was granted to Addison & Stevens, of New York, for a "traveler" or wire loop, sliding around on a single ring; and from this the present form of ring-spinning has been derived. The germ of the idea was undoubtedly in Danforth's "cap"; but the ring and traveler had so much more scope and took so much less power, that it became the favorite, and with the latest improvements seems likely to take the place of the mule for spinning "filling" as well as warp-yarns. Mr. William Mason, of Taunton, Mass., writes me as follows: "I introduced the ring-spinning frame in Connecticut (at Killingly) in 1833. I built quite a number there, but moved to Taunton with all my patterns in the spring of 1836. All the successful ring-frames that were built were made by me up to about 1840, when P. Whitin & Sons commenced to build them. I built ring-frames for their mills as early as 1833. There has been no improvement in the ring and traveler since I first reduced it to practice in 1833."

This system was finally adopted by all the large machine-makers in the United States, and has recently received great developments, of which notice will be taken in due order.

Among other noticeable events of this year (1828) are the establishment of the Covington Cotton Factory at Covington, Ky., of mills at Vincennes, Ind., and of the Norwich Water-Power Company, on the Shetucket River in Connecticut, at what is now called Greeneville.

A substantial stone dam 280 feet in length was built, and several large cotton mills were erected in the few years next following. The Columbian Mills at Mason Village, N. H., also started this year (1828).

In 1829 another roving frame was invented by Gilbert Brewster, of Poughkeepsie, N. Y., in which a temporary twist was given to the roving during its passage from the rolls to the spool, by passing it between two leather bands or belts moving rapidly in opposite directions. This was used for a considerable time to some extent, on account of the small cost of the machine and the great quantity of work it would produce, and was known as the "Eclipse speeder." It was introduced into England by Messrs. Sharp Brothers in 1835, but has now given place to the roving frame with the "equation-box" or "compound" movement, as it is generally called, either in the form of the "fly frame" or "speeder," the latter name being given to those frames in which the arms of the flier are connected at the bottom and are independent of the spindle.

In 1830 the quantity of cotton goods manufactured in the United States was estimated at 250,000,000 yards. The Exeter Cotton Factory, at Exeter, N. H., went into operation in March of this year. The mill at Saco, Me., was burned, and the Lonsdale Company commenced operations at Smithfield, R. I.

In 1831 the York Manufacturing Company, of Saco, secured the services of Mr. Samuel Batchelder, who had just left the Hamilton Company, at Lowell, and built a new four-story mill on the old site. Here, in 1832, Mr. Batchelder introduced the stop motion, which he applied to the drawing frame; and not long after invented the "balance dynamometer," for weighing the power required to drive machinery. Mr. Batchelder remained here until 1846, when he sold out his interest, but some years later, at upward of seventy years of age, accepted the treasurership of the company, at a time when the value of the stock had depreciated from $1,200 to $600 per share, and remained in that office until the value of the shares had again risen to $1,700, when he finally retired from business.

A Convention of the Friends of American Industry was held in New York in 1831, at which the following statistics were obtained: The cotton crop was estimated to be 1,038,847 bales, and the domestic consumption to be more than one fifth of the crop; and the condition of the manufacture in the twelve Eastern and Middle States, including Maryland and Virginia, was as follows:

Capital invested (principally in fixtures)	$40,614,984
Number of spindles in operation	1,246,503
Hands employed	62,157
Value of product, annual	$26,000,000

A large steam cotton factory was built this year at Olneyville, R. I., and another at Fall River, Mass., and the first cotton mill was built at Amoskeag Falls, now Manchester, N. H.

Besides the larger establishments noted, others were growing up on all the waterfalls in Rhode Island, Southern Massachusetts, and Connecticut, the precise data of which I am unable to get. The whole valley of the Blackstone River, from Worcester to Providence, and those of the Mumford, Quinebaug, and Shetucket, furnished numerous mill sites, which were rapidly appropriated, for cotton and woolen factories. Eighteen hundred and thirty-two will never be forgotten as the "nullification year," the Legislature of South Carolina having passed a resolution declaring the tariff act of that year "null and void." The Southern States, who were at first advocates of protection, on account of their "cotton," while New England was generally opposed to it, on account of her "commerce," had changed front with their old adversaries entirely ; and, while the Eastern and Middle States, having embarked heavily in manufactures, were desirous of continuance of the system, the South had begun to think it advisable to ship their cotton to Europe, and under free trade have it manufactured there and returned to her, with such other manufactured articles as might balance the account. The excitement and disturbance consequent on this action of the State of South Carolina led to the introduction and passage of the celebrated "Compromise Act" of 1833, by which the duties on imports were to be gradually reduced.

The statistics for 1833 show the capital invested at Lowell to have been $6,150,000, with 19 mills, 84,000 spindles, 3,000 looms, 5,000 laborers, using 200,000 bales of cotton.

Fall River contained 13 mills, with 31,500 spindles and 1,050 looms, employing 1,276 hands. The York Manufacturing Company started their new mill this year, with 8,000 spindles.

The cotton crop of the United States was estimated the next year, 1834, at 460,000,000 lbs., of which 384,000,000 were exported. At this time, says Dr. Bishop, "the ports of Rio de Janeiro, Aux Cayes, Malta, Smyrna, and the Cape of Good Hope were overstocked with unbleached American cotton, to the exclusion of British goods, which they undersold." I also quote from the same authority, that "the manufacture of cotton gins was commenced on an extensive scale at this time in Autauga County, Ala., by Daniel Pratt, a native of New Hampshire, who in 1846 added a cotton factory. The Jackson Manufacturing Company, at Nashua, N. H., was incorporated this year (1834), and the Lonsdale Company, of Rhode Island, commenced operations.

Eighteen hundred and thirty-five shows 112 cotton mills in the

State of New York, with 157,316 spindles and a capital of $3,669,500, employing 12,954 hands and producing 21,000,000 yards of cloth. The year is also marked by the death of Samuel Slater on the 20th of April. Mr. Samuel Batchelder obtained a patent this year for the application of a steam-drying cylinder to the dressing machine, and the Boston and Lowell Railroad was opened for communication between those two cities. In 1836 the capital invested in manufacturing in the United States was estimated at $80,000,000, the consumption of cotton at 100,000,000 lbs., and the number of spindles at 1,750,000, the value of the product being from $45,000,000 to $50,000,000. The speed of spindles had also been materially increased by the various improvements which had been made, making the production per spindle much greater than in 1813.

This year the Patent Office was organized into an independent bureau, and the Hon. Henry L. Ellsworth was appointed the first Commissioner of Patents on the 4th of July. On the same date a cotton factory of 1,000 spindles was put in operation at Fayetteville, N. C.

Another manufacturing company was incorporated this year at Chicopee, Mass., with a capital of $400,000, under the name of the "Perkins Manufacturing Company"; and the Mauchaug Manufacturing Company, at Sutton, Mass., and the Fiskdale Manufacturing Company, at Sturbridge, Mass., were incorporated. In 1836 the Gray Manufacturing Company, and in 1837 the Laurel Manufacturing Company, of Maryland, commenced operations.

Bishop (vol. ii., p. 411) says : "There were at this date four cotton mills in North Carolina, viz., at Greensborough, Mocksville, Haw River, and Cane Creek. Two or three spinning factories, of 100 or 200 spindles each, carried by animal power, were in operation in Illinois, producing yarn successfully from material grown in the State." One or two manufacturing companies were chartered this year in the State of New York.

In 1838 the Bartlett Steam Mills, at Newburyport, Mass., were chartered. Mr. Bartlett, the chief owner and projector, conceived the idea that a cotton factory in that town would give employment to the families of the sailors and fishermen who were residents of the town, and who were themselves absent the greater part of the year.

The transfer of commerce to the cities of New York and Boston had seriously crippled the prosperity of Newburyport, Salem, and other towns along the coast of New England, and it was hoped that the introduction of manufactures would aid in restoring them to their former prosperity. These towns contained a large unemployed female population, and it was believed that the abundance of labor and the saving in freight of coal and cotton would compensate for the additional cost of steam-power over that of water; but these hopes have proved fallacious, so far as all the steam mills on the seacoast northeast of Boston are concerned. The "James" Mill, at Newburyport, started in 1843, the "Globe" Mill, at the same place, in 1846, and the Portsmouth Steam Mill, at Portsmouth, N. H., the same year, are notable examples.

I will endeavor in an appendix to give some data as to the actual relative cost of the two modes of obtaining power.

The Amoskeag Manufacturing Company, of Manchester, N. H., incorporated in 1831, and which had acquired by purchase all the water rights on Merrimac River, from Nashua to Concord, commenced in 1838 to greatly extend and develop their water-power, by the construction of an extensive system of canals on the east side of the river, opposite to the site of the original mill built in 1810, and erected a large shop for the construction of machinery. In 1839 the Stark Mills and the Manchester Print Works were incorporated, the former commencing operations at once, although the latter did not begin till 1845.

In 1841 and '42 the Amoskeag Company built two mills, which were known as the "Amoskeag New Mills" until the destruction of the old mill by fire some years later, and have since gradually increased to 135,000 spindles, besides building machinery of various kinds and further developing their water-power. The present statistics of Manchester will be given at the close of this memoir, in connection with those of Lowell, Lawrence, Lewiston, and other distinctively manufacturing towns.

Another important event of the year 1838 was the invention by Erastus B. Bigelow of an improvement on the loom for weaving knotted counterpanes.

The close of the decade in 1840 shows a capital of over $50,000,000 invested in cotton manufactures, and 2,285,000 spindles in operation, working 113,059,000 pounds of cotton, and with over 70,000 people employed in its various branches.

The introduction of the self-acting mule, which occurred at this time, is an event of sufficient consequence to deserve a new chapter.

CHAPTER IX.

I HAVE spent a long time in endeavoring to trace accurately the history of the introduction of the self-acting mule into America, and met with some conflicting statements, from which, however, I am enabled to extract the following facts :

Mr. Ira Gay, of the firm of Pitcher & Gay, of Pawtucket, R. I., went to Nashua, N. H., as the mechanical superintendent of the Nashua Company, in 1824. A self-acting mule was then built by him in 1825–'26 at the Nashua Mills, and others of the same pattern were built by his successors, Pitcher & Brown, in 1828, for Edward Walcott, of Pawtucket. "Messrs. Pitcher & Brown afterward built the 'Sharp & Roberts' mule in 1840, and in 1868 built the 'Parr, Curtis & Madely' pattern." For this information I am indebted to the Hon. Zachariah Allen, late Governor of Rhode Island, and through him to Mr. James Brown, of Pawtucket, one of the firm of Pitcher & Brown.

This information as to the early attempts at mule spinning in the United States is confirmed by William A. Burke, Esq., Treasurer of the Lowell Machine Shop, who writes me as follows :

"Mr. Ira Gay was the superintendent of the Nashua Manufacturing Company's machine shop, at Nashua, N. H., when I began to learn my trade in December, 1826. The job I was working on had no part of the mule work, but I recollect very well that they were a prominent machine in the shop, and caused no little discussion with us boys.

"I believe the old Nashua Mill No. 2 was supplied with them for making filling, and I remember very well the tin quill or bobbin, with its wooden cone on the lower end, and the raised rings on the tin part, for holding the yarn from pulling off. As you may now well suppose, these tin tubes were giving trouble enough in keeping straight, for they reached nearly to the top of the spindle. I left off working in the Nashua Company's shop in 1828, and the mules were in a few years abandoned and broken up."

The next attempt was made by William Mason, of Taunton, who writes me as follows :

"I commenced to experiment on the self-actor mule in 1837, and from that time to 1843 I continued to persevere and improve it. I produced several patterns which were more or less successful, but the last pattern, the one that we are now building, was put in operation for the first time at the old Essex Mill, at Newburyport, Mass., early in 1843."

Mr. Thomas J. Hill, of Providence, says : " I made mules after Mason's patent in 1840."

Mr. Mason has, however, introduced many improvements in strengthening the parts of his mules since 1843, and the "Mason mule" is now well known throughout the country, as doing more work with the same power than any other mule built.

Up to 1840, however, mule spinning in this country was only an experiment, and its successful practical introduction was due to Bradford Durfee and William C. Davol, of Fall River, who, on March 9, 1839, signed a contract with Messrs. Sharp, Roberts & Co., of Manchester, England, for the mutual benefit of the aforesaid parties, and William C. Davol acted as agent to procure letters patent for the United States, under the name of Richard Roberts, the original inventor of the mule. These letters patent are dated October 11, 1841.

Owing to the rigidness of the English export laws at that time, the first mule was obliged to be shipped *via* France, and was received at Fall River in 1840. This mule was set up and put in operation the same year at the Annawan Mill, Fall River, under the superintendence of William C. Davol. The mule being considered an entire success, several orders were at once given to Hawes, Marvel & Davol to build them, and the firm, being anxious to complete the orders as rapidly as possible, employed Messrs. Pitcher & Brown, of Pawtucket, who had been experienced in building jacks, to make the carriages from patterns furnished by Messrs. Hawes, Marvel & Davol.

From this date forward the progress of mule spinning in the "Southern district" of Massachusetts and Rhode Island was very rapid, though it was many years before it was introduced to any great extent farther to the North and East, where the prejudice was strong in favor of the hard-twisted and wiry yarn made by the throstle.

Many experiments were, however, made, the first one being the introduction of the Smith or "Scotch" mule by the Matteawan Company, of Fishkill Landing, N. Y., soon after 1840, and letters patent were secured by W. B. Leonard, agent of that company. This pattern was much liked for a time, and a small number of them were built by the Amoskeag Company for a mill owned by them at Hooksett, N. H., and later for one of their own mills at Manchester, and also by the machine shop of the Locks and Canals Company, at Lowell, for a mill at Great Falls, N. H. In 1844 the "Parr, Curtis & Madely" mule, an outgrowth from the Sharp & Roberts, was imported for a mill at Spring Gardens, in Philadelphia, and in 1845 the Franklin Foundry Company, of Providence, commenced their construction.

The Potter mule was also introduced in 1845 by the Manchester Print Works, then going into operation, for the purpose of spinning delaine filling, to which the "Smith" mules were applied at Hooksett; and were afterward imported for the mill built for the manufacture of lawns at Portsmouth, N. H., in 1848, and about the same time for a mill in East Greenwich, Conn.

Still later, about 1853, the "Higgins" or "low-head" mule was introduced by the Franklin Foundry Company, and became well known in the country under their name, and was soon afterward adopted and built by the Saco Water-Power Company.

This mule, the original "Sharp & Roberts" pattern, and the "Mason," were the ones mainly used in the United States up to the period immediately succeeding the civil war, when the extraordinary demand for fabrics, and the impossibility of procuring machinery except at an exorbitant cost, if at all, from any American machine-builder, led to the importation of a great number of English mules of two kinds nominally: one, the "Parr, Curtis & Madely" before mentioned; the other, the "Platt," built by the Platt Brothers, of Oldham, Lancashire; but both, in all essential features, lineal descendants of the Sharp & Roberts mule.

Of these two patterns, that of the Platt Brothers was adopted by the Lowell Machine Shop, and the Parr & Curtis by the Saco Water-Power Company, both of which establishments have been largely engaged in their manufacture since the close of the war; and these, with the Mason and a few of the Sharp & Roberts pattern, are the only kinds now built; the Smith & Potter being quite obsolete, and the Higgins, though an excellent machine, not able to compete in speed or production, on coarse numbers, such as are generally spun in this country, with these improved and more powerful rivals.

Another American mule, invented by Wanton Rouse, of Taunton, which formed the cop by an enormous eccentric cam or "builder," was also introduced in 1853, but has never been used to any great extent.

The English fly frame or roving frame, differing from the American speeder in having the flier attached to the top of the spindle, and revolving with it, while the bobbin was headless, and carried independently by the differential motion of Aza Arnold, already spoken of, which was adopted in England by Houldsworth, was introduced in this country about 1845, and has since been generally adopted, though the prejudice against it, on account of the delay in "doffing," was for a time very strong. The first frames were sent to Rhode Island by Messrs. William Higgins & Sons, of Manchester, who have supplied a great number to American mills, as have also Messrs. Howard & Ballough, of Accrington, particularly during the period following the war; but the manufacture of these machines was soon taken up by the Providence Machine Company and the Saco Water-Power Company and others, and there are now very few imported.

The Lowell Machine-Shop has built an improved speeder, combining the headless bobbin and wind of the fly frame with the long flyer, which is in use to a great extent, and with entire satisfaction, on coarse rovings.

The year 1844 is memorable for the introduction of the turbine wheel, one of which, of seventy-five horse-power, after the Fourneyron plan, with improvements, was introduced at the Appleton Mills at Lowell by Uriah A. Boyden, an eminent engineer of Boston. Attention had been previously called to this matter, and Mr. Elwood Morris, of Philadelphia, had in 1843 published a translation of a French work on the subject of turbines, by Morin, with notes of the operation of some turbines of his own design at Philadelphia; but the success of the system may be said to date from the results obtained by Mr. Boyden at Lowell; seventy-eight per cent. of the gross power of the water, besides that required for driving the bevel-gears and "jackshaft," having been obtained on the test of the first wheel, and eighty-eight per cent. at the test of more perfectly constructed wheels, built afterward from the designs of Mr. Boyden. From this time forward the turbine in some form or another has been introduced, till it has now entirely superseded the old "breast" or "overshot" wheel, giving a much higher percentage of effect from the water, and enabling millowners to run some portion of their machinery in times of freshets or back-water, when the old wheels were entirely useless.

Another American invention of about the same date was the clothshear or trimmer, which is now in universal use in cotton mills, although the intention of its inventor, Milton D. Whipple, of Lowell, was confined to trimming the ends and threads from the cloth in the calico print works; but the machine proved so useful that it has been generally adopted in all mills as a necessary operation in preparing the cloth for market.

The growth of the cotton manufacture, under the stimulating influences of protection, and the favor with which American goods were received in China, was now very rapid, and in 1845 plans were made for a further development of the water-power of the Merrimac River, at North Andover, by the construction of a dam across it, at the rapids at that place, which should give a fall of twenty-six feet, and set back the water in the pond above the dam to the foot of Hunt's Falls, just below Lowell. The Essex Company was incorporated for that purpose, and the work commenced; and in 1847 the dam and canal were completed, and the town which had sprung up in consequence of the operations was called Lawrence, from the name of the gentlemen in Boston who had been the leaders in the enterprise.

The Atlantic Cotton Mills were the first to commence operations, but were soon followed by others, until a flourishing and populous city occupies the site which thirty years ago was a barren sand-bank, and of which particulars will be given in the proper appendix.

The Dwight Manufacturing Company, of Chicopee, had been incorporated in 1841, and eventually absorbed both the Cabot and Perkins Mills, of the same place.

Another special industry was inaugurated by E. B. Bigelow, of Lancaster, in 1844, by the commencement of the Lancaster Mills, at Clinton, Mass., for the manufacture of ginghams, which were to be woven by machinery instead of by hand loom, as had formerly been the practice, and resulted in entire success, proving the scheme to be practicable and profitable, and serving as the pioneer to various other successful establishments of a similar character.

The Ocean Mills, of Newburyport, Mass., were commenced in 1845, the Boston Duck Company, of Palmer, in 1843, and the Plymouth Cotton Company in the same year; and in 1847 the Wamsutta Mills, of New Bedford, the Agawam Canal Company, of West Springfield, and the Annisquam Mills, of Rockport, all in Massachusetts, were started. In 1848 the Glasgow Company, of South Hadley, followed the Lancaster Mills on ginghams. The Massasoit and Metacomet Mills, of Fall River, commenced respectively in 1845 and 1846, and the Naumkeag Mills, of Salem, commenced in 1839, and the Otis Manufacturing Company, of Ware, 1840, should also be included in the growth of Massachusetts for this decade.

Nor were the other New England States behind in developing their resources. The Saco Water-Power Company, in Maine, by means of new dams and canals on the west side of the river, opposite the York Mills, utilized the whole power of the Saco River, and built the Pepperell Mills in 1844, and the Laconia Mills in 1845, with a large machine shop, which has been successfully operated since by the Water-Power Company, and as a necessary result the city of Biddeford grew up around these establishments. The Hallowell Manufacturing Company, at Hallowell, was also commenced in 1845.

Neither can the growth of the cotton manufacture at this time be estimated fairly by the number of mills built. The original mills of 4,000 to 6,000 spindles had given place to larger and more convenient structures containing from 10,000 to 15,000 spindles each, and these were in time to give place to still larger ones, or to be connected by intermediate buildings, bringing 30,000 or 40,000 spindles under one roof and one system of superintendence, as experience and practice developed overseers of skill sufficient to take the charge of so large an amount of machinery, with its complement of operatives.

In New Hampshire the Amoskeag and Great Falls companies were extending their operations and adding to their machinery, and in 1845 the Monadnock Mills at Claremont were commenced, using the water-power of Sugar River.

Large numbers of mills were also built in Rhode Island, among which were those of the Groton Company, at Woonsocket, in 1840, the Hope Company, at Scituate, in 1845, the Warren Company, at Warren, in 1847, and the Quidnick Company, at Anthony, in 1848.

In Connecticut the Falls Company, of Norwich, and the Chestnut Hill Mill, at Killingly, commenced operations in 1844; the Granite Mill, at Stafford Springs, and the Greenwood Company, at New Hartford, in 1845; the Uncasville Manufacturing Company, at Montville, in 1848; and the Moodus Manufacturing Company, at East Haddam, in 1849.

The Victory Manufacturing Company, of Saratoga, N. Y., commenced in 1846, and the Utica Steam Mills in 1848.

I have no statistics by which to mark the exact progress of the manufacture at this period in the States farther South and West, but it is certain that there was a large increase in and around Philadelphia, which has always been a great manufacturing center, though the individual enterprises have been on a smaller scale than those of the great incorporated companies of the New England States.

The census of 1850 gives no reliable information as to the amount of machinery then in operation or the number of operatives employed, but the production of cotton fabrics is given as 263,190,642 lbs.

The business of cotton manufacture was by this time so firmly established as to be little affected by changes in legislation in regard to duties on the coarser fabrics required for domestic consumption, to which American machinery had been adapted; and its progress was constant and steady for the next ten years, with constant improvements in the mechanical and economical appliances, and successive enlargements of the scale of operations; and its growth from 1850 to 1860 will fill our next chapter.

CHAPTER X.

The earlier years of the next decade, until 1857, were marked by a continual and steady growth of the cotton manufacture; the American manufacturers and mechanics who visited the first International Exhibition in London in 1851 extended their tours to the manufacturing districts, and brought home many valuable ideas and economical improvements, which were rapidly introduced in all parts of the country. The boldness of the scheme which created a waterfall at Lawrence, by the erection of a dam twenty-five feet in height and the formation of a mill-pond ten miles long, had stimulated an enterprise on a still larger scale—that of rendering available in a similar manner the enormous power of the Connecticut River at South Hadley, where there was a fall of sixty feet extending over some two miles, in a series of continuous rapids. To accomplish this purpose the Hadley Falls Company was incorporated by the Legislature of Massachusetts in 1848, with a capital of $4,000,000, and operations were commenced by the purchase of about 1,200 acres of land on the west side of the river, near the principal fall, where a dam of timber, loaded with stone, with massive stone bulkheads, was constructed, 1,019 feet long and 30 feet high. The first attempt was unsuccessful, and the dam gave way under the pressure of the water as it was being completed; but a second attempt fared better, although the wearing away of the sandstone bed of the river below the fall necessitated a reconstruction of the work in 1868, when an apron was built below the dam, so as to give the whole structure the form of a triangle, with a base of ninety feet and a perpendicular of thirty, consisting of a heavy timber crib frame, bolted to the rock, filled with stone and covered with plank, while the crest was "armor-plated" with boiler iron. In the western bulkhead, operated by a turbine wheel, were placed the gates which admitted to the upper canal, and from this the water was taken to a second, and then to a third, which discharged into the river at the lower end of the town.

The fall from the upper to the middle canal was 20 feet, from the middle one to the lower one 12 feet, and 20 feet to the river at the upper end, where part of the water was discharged, while it was from 23 to 28 feet on the lower one, and the whole system was over four miles in total length, rendering available in all about 30,000 horsepower.

The first dam was completed in 1849, and a machine shop (since converted into a cotton-thread mill) was constructed, in imitation of the original plan of Lowell, where a shop to build the machinery for the mills was an absolute necessity and an integral part of the system, from the entire absence in the country at that time of any shops of sufficient capacity for the purpose. A machine shop (also since converted into a cotton mill) had been built at Lawrence, and it was the first thing done at the new town, now a city, of Holyoke.

In 1852 the Hampden Mill, of 16,000 spindles, was built, and in 1853 the Lyman Mills, now containing 75,000 spindles, commenced operations. The crash of 1857 ruined the original company which built the dam and laid out the town, and the water-power passed into the possession of the Holyoke Water-Power Company, and has since been largely utilized for the manufacture of paper and other industries, Holyoke being now the great headquarters of the writing-paper business.

At about the same date another company commenced the development of the water-power of the Androscoggin River at Lewiston, Me., where a high fall and a solid rock foundation rendered operations much less expensive, and where manufacturing has been very successful.

The first cotton mills to start here were those of the Bates Manufacturing Company, in 1852, since followed by others, the statistics of which will be given at length later in this memoir.

In addition to these larger enterprises, the following establishments commenced operations between 1850 and 1860 : In Maine, the Cabot Manufacturing Company, of 35,000 spindles, at Brunswick, in 1857, and the Westbrook Manufacturing Company, at Saccarapa, 16,000 spindles, in 1858. In Massachusetts, the mills of the Hebron Manufacturing Company, at Attleboro, 37,000 spindles, in 1852; the Monument Mills, at Great Barrington, the same year; the Lawrence Duck Company, 7,500 spindles, at Lawrence, the Phœnix Company, 6,000 spindles, at Shirley, and the American Linen Company, at Fall River, originally projected for a flax mill, but changed to cotton, now containing 83,000 spindles, in 1853; the Warren Cotton Mills, 13,500 spindles, at West Warren, and the Ward Manufacturing Company, since changed to the Indian Orchard Mills, 16,000 spindles, at Springfield, in 1854—all on cotton exclusively, besides the Pacific Mills at

Lawrence, started in 1852 partially on cotton and worsted goods, as well as on calicoes, originally on a magnificent scale, and now one of the largest manufacturing establishments in the world.

In Rhode Island during the same time were commenced the mills of the Franklin Company, at Olneyville, in 1850, 34,500 spindles ; the Valley Falls Company, at Lincoln, 35,000 spindles, and the Lippitt Company, 7,500 spindles, at Phœnix, in 1853 ; the Dyerville Company, 18,000 spindles, at Centredale, and the Clinton Mill, 16,000 spindles, at Woonsocket, in 1854 ; together with the Social Mills, since burnt and rebuilt, now 50,000 spindles, at Woonsocket, in 1855.

Connecticut also shows a long list of mills, commencing with the Quinebaug Company, 34,000 spindles, at West Killingly, in 1851 ; the Smithville Manufacturing Company, at Willimantic, 17,000 spindles, and the Atlantic Duck Company, at Haddam, the same year ; the East Haddam Duck Company and the Wauregan Mills, now containing 56,000 spindles, were begun in 1853 ; and the Williams Duck Company, at Haddam, in 1854, the duck manufacturing companies being all small establishments. In 1856 the A. & W. Sprague Manufacturing Company built their great Baltic Mill, of 75,000 spindles, for the manufacture of print cloths, at the village of Sprague ; and in 1857 the Willimantic Linen Company, so called, but devoted to the manufacture of sewing-cotton, commenced operations, and now employs 45,000 spindles on that product. The Dunham Company, of Willimantic, with 6,000 spindles, was started in 1858, and the Williamsville Manufacturing Company, with 12,500 spindles, at West Killingly, and the Elliottville Manufacturing Company, at East Killingly, 3,800 spindles, in 1859. The Attawaugan Company, also at Killingly, 17,000 spindles, began in 1860.

A part, however, of the above named establishments commenced operations with a smaller number of spindles than they are now credited with, having received large additions since the date of starting.

The Newburgh Steam Mills, at Newburgh, N. Y., were commenced in 1850, and the Harmony Mills, at Cohoes, reorganized, as before noted.

The Indiana Cotton Mills, at Cannelton, Ind., with 10,800 spindles, commenced in 1855, and the St. Louis Cotton Factory, at St. Louis, Mo., with 10,500 spindles, in 1857.

The financial difficulties of the latter year checked any further rapid development until after the close of the civil war in 1865, and proved fatal to the great enterprise at Holyoke, causing a great loss of capital to the original projectors, as well as to those of many of the other new establishments which had just commenced operations.

The great mechanical invention of the period was that of the "self-stripping" card, which was brought into successful use about the year 1857. Various experiments had been made both in this country and in Europe to accomplish the object of cleansing the "top-flats" or cards from the constantly accumulating waste, consisting of short cotton, bits of seed-husk, leaf, etc., by some automatic process while the card was in operation, thus avoiding loss of time and securing more systematic, regular, and perfect cleaning, while dispensing with a large amount of hand labor. This result was finally accomplished by two inventors, working separately, but whose patents were finally united in 1867, forming the card as in ordinary use to-day.

George Wellman, of Lowell, Mass., received letters patent December 6, 1853, for a system of mechanism for elevating, cleaning, and returning to their places the top-cards or flats, and also for moving the vibrating frame, which accomplished this purpose from flat to flat, or from one to the second flat from it in the series.

Horace Woodman, of Biddeford, Me., received letters patent on the 1st of August, 1854, for a system of mechanism for the same purpose, but differing somewhat from that of Wellman.

Wellman took out a second patent, for an improved machine, March 18, 1856, and a third one January 27, 1857.

Woodman also received a second patent July 8, 1856, and a third December 1, 1857. As might naturally be expected, where the object to be effected was so positive and confined, these different patents interfered with each other essentially, and, after much litigation, the matter was compromised by a union of the two in 1867, as noted. This invention has come into general use, and has not only proved more effectual in producing perfect work, but more economical in wear and tear of card clothing, as well as in labor.

Another valuable invention, but of less general application, was the seamless bag loom, invented by Cyrus W. Baldwin, of Manchester, N. H., in 1851, in which, after weaving the necessary length of bag in a tubular form, by a simple automatic change of cams, the harnesses were shifted in such a manner as to weave an inch or more of solid double cloth, thus forming the bottom of the bag, and then returned to their original position in the same manner, without stoppage or disarrangement of the machinery or need of manual assistance.

The parallel motion for the picker-staffs of looms was also brought into use about this time, the first one having been invented by W. W. Dutcher, of Hopedale, Mass., in 1853, soon followed by several others. This was a valuable invention, as saving a large expenditure for "pickers" and "picker-strings," and in some form or another is in general use.

The "shuttle-guard" of H. D. Robbins, patented in 1852, and the improved oiler of I. R. Scott, afterward improved by W. H. Thompson, are also deserving of notice.

In connection with the improvement in cards should be mentioned the railway evener, invented by D. W. Hayden, of Willimantic, Conn., in 1850, and afterward improved by Newell Wyllis, of Glastonbury, and still later by Messrs. George Draper & Sons, of Hopedale, Mass. This valuable invention, applied to the railway-head, which receives the sliver from a system of six or more cards, being driven by a belt working on a pair of cone-pulleys, changed the draught of the rolls in the head whenever the sliver was broken down from any of the cards, by means of a "trumpet," through which the combined sliver passed, and which operated a lever, shifting the belt on the pulleys, thus causing the sliver delivered to be of uniform size.

In 1860 the large machine shop at Lawrence, Mass., built by the Essex Company, and afterward known as the Lawrence Machine Shop, was converted into a cotton mill, and has since been known as the Everett Mills; and the Pemberton Mills, at the same place, originally built in 1853, which had been destroyed by an accident, with horrible loss of life, the previous year, were rebuilt; and both these mills were applied to the manufacture of colored fabrics, such as cottonades and dress-goods.

These are the principal enterprises of this period, during which the number of spindles had increased, as shown by the census of 1860, to 5,035,798, and the pounds of cotton worked to 450,877,823, while the crop of cotton raised in 1859 had reached the enormous amount of over 4,000,000 bales, or 1,850,000,000 pounds, of which we consumed about one quarter, while the rest found a ready market at high prices in Europe, and the enormous profits realized by its cultivation gave rise to political results which for a time convulsed the world and bid fair to destroy the republic.

CHAPTER XI.

It is not within the purposes of this article to enter on the discussion of political questions, but this enormous and rapid growth of the cotton manufacture was unquestionably one of the principal causes which brought about the great American Civil War.

The superiority of the staple of American cotton had made it the favorite in the markets of the world, and the demand for it seemed to be practically unlimited. The profits of its cultivation were great, and every acre of the cotton-growing States that could be made available was devoted to this purpose. "Cotton is king" was the watchword of the planters, and so fully were they convinced of the truth of this fallacy that they dreamed of a new empire to be devoted to the cultivation of cotton by slave labor, and to be attained by the disruption of the existing union with the manufacturing and food-growing States of the North and West.

The question of slavery had long been a serious difficulty between the North and the South, and, as it was fully believed in the latter region that this important staple could only be raised by slave labor, the planters regarded any opposition to the system or its extension as an attack upon their most vital interests, while the North, regarding it as both morally and economically wrong and injurious, had planted itself firmly against its introduction into the unorganized territories belonging to the nation.

The election of a president on this basis, of opposition to the further extension of slavery, served the South as a *casus belli*, and, believing that cotton must be had, and that they were sure of the support and assistance of European manufacturers, the cotton-growing States took the responsibility of attempting the dissolution of the Union.

It is needless to say that the result has been very different from the anticipation, and that the crop of cotton cultivated by free labor is now greater than before the war, having reached 4,500,000 bales, and continually increasing, although the whole system has been changed, and the crop is now raised by small land-holders or tenants of land, in connection with food-crops, as a staple "money-article," for the supply of their clothing and other wants beyond their own production, instead of in large lots by the owners of great plantations, who depended on the profits of their cotton to supply all their other wants, even buying much of the food consumed by their families and laborers. The greatest increase in the crop has been in the State of Texas, where the most white labor has been employed, and the "farm system," as in contrast to the "plantation," most fully developed, amounting to over 80 per cent. more than before the war, the last crop of the State having been nearly 700,000 bales.

The introduction of the new system has of course been attended with more or less difficulty to the manufacturer, who can not now go into the market and purchase one or two hundred bales of cotton of the growth of one plantation and of uniform quality; but care on the part of the buyer and skill in mixing his cotton on that of the manufacturer render it one of no very great importance.

The outbreak of the war in 1861 stopped all further extension of the manufacture for a time, and utterly prostrated the business. Many cotton mills sold their stock on hand, and put in more or less woolen machinery, to supply the anticipated demand for army clothing, or in doubt whether they should ever see any more cotton; while other shrewder manufacturers bought the cotton thus sold, and piled it up, until the demand for cotton cloths rendered its manufacture enormously profitable; but it was a long time before matters resumed their normal condition.

The necessities of the South, however, drove them into manufacturing to a small extent, and in 1864 the Augusta Cotton Factory, of 23,000 spindles, was commenced; and since the close of the war the number of spindles at the South has been increasing, although the deficiency of capital and skilled labor prevents very rapid progress. Still Virginia, North and South Carolina, and Georgia have begun to utilize their valuable water-powers to some extent, and something has been done in Alabama, Kentucky, and Mississippi.

After about two years of uncertainty, the manufacturing districts began to take courage, and business was again resumed partially; to be revived with increased vigor at the close of the war in 1865, when the country had become pretty thoroughly drained of its stock of cotton fabrics; and while many of the existing mills increased their number of spindles to a great extent, the following new enterprises of note were commenced—one or two of them during the war, but the greater number after its close. In 1861 the Webster Mills, at Suncook village in Pembroke, N. H., which had been previously planned, were started, with 30,000 spindles, in 1862 the Coventry Company, at Anthony, R. I., with 10,700 spindles, and the Grafton Mills, at Grafton, Mass., with 12,400 spindles; in 1863 the Ashland Company, 20,000 spindles, at Jewett City, Conn.; the Central Mills, 10,600 spindles, at Southbridge, Mass.; and the Oriental Mills, at Providence, R. I., with 15,000 spindles; while the machine shop of the Hadley Falls Company, at Holyoke, was converted into a mill for the manufacture of spool cotton, and reorganized as the Hadley Company, with 30,000 spindles. In 1864 the Providence Steam Mill, at Providence, R. I., began with 21,000 spindles; the Reynolds Manufacturing Company, at Bristol, R. I., with 10,750 spindles; the Augusta Factory, at Augusta, Ga., 23,000 spindles; and the Indianapolis Manufacturing Company, of Indiana, with 4,000 spindles. The year 1865 saw the start of the Whitestone Company, at East Killingly, Conn., with 8,000 spindles; the Danielsonville Company, at West Killingly, 16,000 spindles; the Putnam Mills, at Putnam, Conn., 18,600 spindles; the Merrick Thread Company, at Holyoke, Mass., 12,000 spindles; the Harris Manufacturing Company, Coventry, R. I., 14,000 spindles; the Orion Manufacturing Company, East Greenwich, R. I., 15,000 spindles; the Rockville Manufacturing Company, Rockville, R. I., 8,000 spindles; the United States Flax Manufacturing Company (on cotton goods, however), at Pawtucket, 30,000 spindles; the Fletcher Manufacturing Company, at Providence, 17,000 spindles; and the Warren Manufacturing Company, at Warren, Md., 6,000 spindles. In 1866 the Williston Mills, at East Hampton, Mass., went into operation, with 30,000 spindles; the Williamstown Manufacturing Company, at Williamstown, Mass., 14,600 spindles; and the Empire Manufacturing Company, Paterson, N. J., 3,000 spindles. In 1867 the A. & W. Sprague Manufacturing Company built a large mill at Augusta, Me., making, with the old mill on the spot, which they purchased, a total of 40,000 spindles; and the dam across the Kennebec River at that point was rebuilt, affording a magnificent water-power, which has not as yet been further developed. The Renfrew Manufacturing Company, at South Adams, Mass., with 27,000 spindles, was commenced the same year, and the Smithfield Manufacturing Company, 11,000 spindles, at Hyde Park, Mass. In Vermont the Vermont Mills at Bennington, 7,000 spindles, and the Burlington Cotton Company, at Winooski Falls, were started. Another large Southern enterprise also dates from this year—the Eagle and Phœnix Manufacturing Company, at Columbus, Ga., with 22,000 spindles; and in 1868 the Frankfort Cotton Mills, at Frankfort, Ky., were commenced. The year 1869 records the beginning of the Lowell Hosiery Company, 10,400 spindles, at Lowell, Mass.; the China Mills, 50,000 spindles, at Suncook, N. H.;

the Mt. Vernon Manufacturing Company, at Alexandria, Va., 4,000 spindles; and the Marshall Manufacturing Company, with 6,000 spindles, at Manchester, Va. In 1870 the Slater Cotton Company, at Pawtucket, R. I., 20,000 spindles; the Whitin Manufacturing Company, Northbridge, Mass., 16,000 spindles; the Morse Mills, Putnam, Conn., 10,000 spindles; and the gigantic Ponemah Mill, of 72,000 spindles, at Taftville, in Norwich, Conn., were commenced.

This closes the list of the more important manufacturing establishments commenced during the decade, during which the increase of spindles, as shown by the census of 1870, was very great, and was to continue for three years longer, although, owing to the substitution of lighter fabrics, on account of the scarcity and high price of cotton, the number of pounds worked was slightly less than in 1860, being 447,216,000 pounds, while the spindles had increased to 7,114,000.

There is no American invention of any remarkable importance till the close of this period; but a very great advance had been made by the adoption of the "slasher" dresser, which was introduced from England at the Androscoggin Mills, Lewiston, by Mr. A. D. Lockwood, in 1867. The distinctive feature in this machine consisted in the use of hot size, through which the warp was passed, and then dried at once over large cans or cylinders, made of copper or galvanized iron, instead of being passed through cold starch, usually fermented, and dried by the air as it passed to be wound on the beam for the loom. Previous trials of machines on this principle had been made in this country, one known as the "tape-dressing" machine having been imported by Mr William A. Burke, then superintendent of the Lowell Machine Shop, in 1851; but whether from imperfection in the yarn, or prejudice on the part of the operatives, none of the experiments were successful until the arrival of the machine imported by Mr. Lockwood.

One of these machines, as built by Messrs. Howard & Bullough, of Accrington, Lancashire, proved capable of doing the work of ten of the old style of dressers, or from 300 to 500 pieces per day; while the expense per yard was only about one fifth of that of the old manner, and the warp was more thoroughly sized, so that the new "slashers," as they were called, were soon introduced in every direction.

There had been many minor improvements in warpers, spoolers, and looms, but spinning had remained without any essential change since the introduction of the ring spindle, in 1836, by Wm. Mason, and of the self-acting mule in 1840, by the importation of the Sharp & Roberts mule, and the invention by Mr. Wm. Mason, of an essentially different machine in construction, but capable of accomplishing the same purpose, about the same time. The ring spindle had taken the place of the flier throstle, in nearly all the new mills, and with a saving of 20 per cent. of the power, but had not been materially changed from its first form, unless to make it heavier so as to run steadier, until 1870, when a great alteration was made in it by Oliver Pearl, of Lawrence, Mass., producing very important results and materially reducing the cost of production. This may perhaps be best understood by Mr. Pearl's own description, in his patent issued May 3, 1870, viz. :

"My invention relates, *first*, to certain improvements in the construction of bobbins having frictional or adhesive bearings uniting them to the spindle and carried by it, the object of this part of my invention being to make a very light bobbin, and strengthen its various parts so that it will not be easily crushed or broken ; *second*, to an improved construction, and combination of both the bobbin and the ring-spindle, so that they can be successfully used with greater advantages of length of traverse, speed, and steadiness of rotation than heretofore attained, and at the same time be much lighter, the object of this part of my invention being to greatly diminish the amount of power required to drive the spindle at any given speed, and increase its efficient operation at the same time."

After describing the ordinary spindle and bobbin, he says of his own : "This bobbin is made with a thin and light shell or barrel of wood, and has a lower adhesive or frictional bearing, k, and a middle one, i, and is also bushed at the upper end by a plug, re-inforce, or bushing, l; and the bearings, k and i, and the bushing, l, are united to and combined with the shell of the bobbin, and strengthen it in all directions from being broken. The adhesive or frictional bearings, k and i, are made to sustain the bobbin on the spindle in one position with relation to the latter, and so as to enable the spindle to carry the bobbin with it in its rotation.

"My improved spindle, instead of extending substantially to the upper end of the bobbin, as heretofore, is only made long enough above the upper bolster to enable the adhesive bearing i at the center of the bobbin to hold the latter firmly upon it. I am thus enabled to remove a large portion of the blade of the spindle above the bolster; and the tube of the bobbin projecting beyond the shortened blade of the spindle, resting, by its adhesive central bearing, upon the latter, and being both light and rigid, retains its length and the position which it had before the spindle blade was shortened, while the traverse of the frame and the length of the bobbin remain as before.

"By thus dispensing with the length and weight at the top of the spindle above the bolster, while the length of bobbin and the traverse of the frame remain as before relatively, I am enabled to lighten the lower part of the spindle and wheel below the bolster many times the weight taken from its blade above, without destroying the proper balance of the spindle and its consequent steadiness of motion ; and by these means I accomplish the ultimate effect, which is the purpose of this improvement, of enabling the spindle to be run steadily at high speed with much less power than heretofore, thus diminishing the expense and increasing the production at the same time."

I have been thus particular in giving Mr. Pearl's own description of his invention, as he was soon followed by others, who aimed at accomplishing the same result by different methods, but the claims of some of whom, infringing more or less on those of Mr. Pearl, are now the subject of legal adjudication. The effect of the improvement was decided and positive, the new spindle ran at the same speed with the old one, with a saving of one third of the power required to carry it, and was also capable of being run at a higher speed than had been possible before this invention.

The old spindle was limited to a speed of about 6,000 revolutions per minute, and was seldom used at even that speed, probably averaging about 5,500 revolutions in ordinary practice, while the new one was capable of being driven to 6,500 or even to 7,000 revolutions without extra vibration or loss of power.

Part of this saving was due to the diminution in weight of the spindle, which had been reduced from 12 ozs. to about 5 ozs., but more to its smaller diameter, giving a shorter length of frictional surface in the bearings, opposed to the lateral tension of the driving band ; but a great deal of it was due to the improvement in the bobbin, which was reduced from 1 oz. or $1\frac{1}{4}$ oz. in weight to about $\frac{1}{2}$ oz.; thus, with the shortening of the top of the spindle, materially reducing the *vibratory* weight, carried above the upper bearing or "bolster."

The new bobbin was more expensive than the old one, but, from the peculiarities of its construction, was necessarily more accurately and carefully made, and more mathematically "true," and less liable to warp, and spring from its form of a true cylinder.

As, according to the mathematical law, the vibration of a spindle is as the cube of its length above the point of support, divided by the cube of its diameter, this disturbing element, which caused much of the friction, and thereby absorbed a large proportion of the power, was materially reduced by the new invention.

Mr. Pearl commenced his experiments on this spindle and bobbin in 1868, but did not take out his patent until 1870.

He was soon followed by Jacob H. Sawyer, of Lowell, who, on the 11th of April, 1871, received letters patent for an improved spindle, which was a further step in carrying out the same mechanical law, by placing his upper bearing or "bolster" at the top of a tube, supported from the "bolster-rail," and chambering out the lower half of the bobbin, so that it would drop freely over this tube; thus placing the point of support very nearly in the center of the bobbin vertically, and still further diminishing the length, and consequent vibration, above the point of support.

In both cases, the "blade" of the spindle or that part above the bolster was slightly tapered, and in Mr. Pearl's case the bobbin was driven by the adhesion to it of two bushings, one at the bottom, and one half way up the bobbin, while in Mr. Sawyer's spindle the bushings were respectively at the center and the top, or entirely in the upper half of the bobbin, which in Mr. Pearl's invention projected above the top of the spindle.

A difficulty was soon found with the Sawyer spindle, from the inability to oil the bolster bearing while in operation, which was obviated by Mr. George Draper—who purchased the patent, or a part of it—by the application of a bolster tube having a spiral groove cut internally, so as to carry up the oil steadily and constantly from a supply at the bottom, to the bearing at the top, while the spindle was in revolution. This spiral bolster was patented January 14, 1873, and proved successful.

The natural result of Mr. Sawyer's invention was to make a still further increase in the capability of speed attainable by the spindle, and reduction of the power required, over that reached by Mr. Pearl, although the spindle, including its bearings, was necessarily more expensive in construction; but both forms are being very extensively introduced, as might be expected, when we consider that one half, or nearly so, of the whole power of a cotton mill was required for the spinning, and that of this power one third was saved over the old style of Ring spindle, and one half over the Flier spindle by their adoption, the saving being rather greater with the Sawyer spindle than with the Pearl spindle, for the reason that its diameter is usually less.

Another form of light spindle, invented by Richard Garsed, of Philadelphia, was introduced by the Bridesburg Manufacturing Company, in 1872. It consisted like the others of a short spindle, but the bobbin had no adhesive bearings, being driven by a clutch at the bottom, which engaged on a square shoulder cut on the spindle just above the holster; and, the spindle being straight, the bobbin was so bored as just to drop freely on it to its bearings.

Another feature about this spindle was the use of a "loose bolster," patented by Barton H. Jenks, of Philadelphia, which was held by a screw from turning around in the "rail," but had just enough play to allow it to align itself to the spindle.

This not only prevented the spindle from being cramped between the upper and lower bearings, but avoided the friction caused by the revolution of a body liable to vibration, in a rigid bearing, to some extent. This form of spindle, although not possessing the advantages of the Sawyer plan, where a very large and heavy bobbin of yarn was to be carried, has proved itself very useful with light bobbins, especially for weft or filling, as there can be no loss of twist in the yarn, from the bobbins working loose upon the spindle, and as the size of the weft bobbin is limited by the capacity of the shuttle. The saving of power within the range of its capacity is about the same as with the Sawyer spindle.

Still another pattern, the "Rabbeth" spindle, is a modification of a spindle patented in 1867 by Messrs. Rabbeth & Atwood, of Birmingham, Conn.

The original spindle was intended for use in the manufacture of silk, and had a sleeve or shield, extending downward below the flange on which the bobbin was set, and by means of pins in which it was driven, for the purpose of protecting the silk from any oil which might be thrown out from the bolster by the revolution of the spindle.

The spindle itself was dropped into a tube filled with oil, at the bottom of which was the step, or lower bearing. The spindle thus revolved in oil, which was prevented from getting on the silk by the sleeve or shield above mentioned. As this tube prevented the application of the whorl to the spindle in the usual manner, it was attached to the bottom of the sleeve.

This form of spindle was modified by Mr. Rabbeth in 1872, by the use of a bolster similar to Mr. Sawyer's, and the reduction of the sleeve in diameter, so as to permit the chambered bobbin to drop down upon and be driven by it.

This brings the bobbin, spindle, and bolster in relatively the same mechanical position as in the Sawyer patent, but differs in having the lower part of the spindle constantly immersed in and revolving in oil, thus needing less frequent attention. This spindle has also been introduced to a considerable extent, with nearly the same results in economy of power as the others.

Several other light spindles have been experimented with, but have proved to be an infringement on one or the other of the above forms, the rights of which as among themselves have not yet been entirely adjudicated, but of which in all almost 2,000,000 have been put in operation, the Sawyer spindle so far showing the greatest number.

The saving of power resulting from their use has proved of great value to mills, where steam was employed for that purpose, as well as to those driven by water, on the many privileges where the supply was short during the summer months.

The subject of opening and picking cotton also received much attention about this time. Various machines had been employed for this purpose, the most of them of English origin, prominent among which was the one known as the Creighton Willow, which opened the cotton very successfully but delivered it in loose masses, on the floor, and was very liable to accident from fire.

Messrs. Lord, of Todmorden, England, and others, and Kitson, of Lowell, Mass., turned their attention to machines which should not only open the cotton from the bale, but form it at one operation into a preparatory lap, for the second machine. These machines came into general use, on account of the thorough way in which they did the work, and their immunity from fire, but were objectionable from the great amount of power required to operate them, and were also supposed by some to injure the staple of the cotton.

This form was originally introduced in England during our Civil War, when the English spinners had to depend largely on short-stapled and dirty East Indian cotton, and answered their purpose admirably, but did not seem to be so well suited to the longer stapled American varieties. I have referred to their operation in the supplementary chapter, in the notice of Kitson's improvements. On the 20th of December, 1870, an improved opener was patented by Messrs. Palmer & Jillson of Willimantic, Conn., in which the rigid beater blade was replaced by a series of hinged fingers, hung on rods parallel with the axis of the beater, and which, while striking the cotton with a blow like a flail, as it was delivered by the feed rolls, would yield partially to the resistance of an unusually heavy clot or mass due to dampness in the cotton, and great pressure in the packing. There were, however, objections to the operation of this machine, as the points of the fingers which struck the cotton tended to draw it out into "strings," so called, and this difficulty was obviated by the improvement of Messrs. Whitehead & Atherton, of Lowell. Patented in 1871, 1872, 1873, 1874 and 1875.

This improvement consisted in uniting the ends of every pair of fingers by a cross-bar, thus forming a continuous line of short beaters, each about five inches long, and hinged on the above-mentioned rods. This "Whipper" as it was called avoided the objection to the Jillson & Palmer "finger," cleaned the cotton from seeds as thoroughly as the rigid beater, or even more so, required much less power in operation, and caused less injury to the staple. This machine of Messrs.

Whitehead & Atherton has been very widely and successfully introduced, and has led to the improvements already mentioned by Mr. Kitson, in which the lumps or masses of cotton are torn apart or loosened before reaching the beater, whose office it is to remove the seed. Both the improved opener of Mr. Kitson and the Whipper opener of Whitehead & Atherton are now being introduced in England, and giving entire satisfaction, the Whitehead & Atherton machine having proved itself especially adapted to long-stapled cotton, or to cotton which has been previously dyed.

Another variation introduced by Kitson was the substitution of a series of toothed feed rolls, increasing in speed, like those in a drawing frame, by which the cotton was torn apart, and then blown through a dust-trunk, before reaching the regular feed-rolls of the beater.

In either way, the improvement in the manner of opening cotton from the bale has been very great.

The city of Fall River made its great increase in manufacturing at this period, and a number of new and large mills were erected, almost entirely devoted to the manufacture of print-cloths, the demand for which was very great, and the price of which had risen to 8 cts. per yard, or 50 cts. per lb., giving a very large profit. These mills were, to some extent, filled with English machinery, the roving frames, mules and dressers being generally imported, but the cards and looms were of American manufacture. A reference to the table of statistics of Fall River will show the date at which these new mills were built, and the rapidity of their increase, which, with that of other places, resulted in 1873 in overstocking the home market with cotton goods. Another result also followed from the large increase in mule spindles.

These machines were usually accompanied by foreign operatives, who brought with them all their foreign prejudices, as well as their skill at their trade, and soon attempted, by trades-union management, to fix the price of labor and dictate to the mill owners, by means of a "strike"; in which, as usual, the operatives came out the losers, after stopping the mills for many weeks. The labor of these weeks was not only lost, but the attention of manufacturers has been turned to the production of weft as well as warp yarns, by the improved light ring spindle instead of the mule. This is now the subject of experiment by various inventors, and has nearly passed the stage where it can be called experimental. It has been satisfactorily proved that a soft weft yarn can be spun on either the Sawyer, Pearl, Garsed, or Rabbeth spindles, at as low a cost as on the mule, by a more docile and manageable class of operatives, and with the advantage of producing an equal amount of yarn, with one half the quantity of room in the mill. Many large establishments are trying one or more of these different spindles, and it seems very probable that their substitution for the mule will be gradually effected, when the best form of spindle is decided upon.

Besides the great increase at Fall River, the following manufacturing establishments of note commenced operations during this period, 1871, in Massachusetts. The Potomska Mill at New Bedford, 44,000 spindles; the Freeman Manufacturing Company, 17,000 spindles, at North Adams, in 1874; and the Johnson Company at the same place with 6,000 spindles. In Connecticut, in 1872, the Powhattan Manufacturing Company, 17,000 spindles, and the Mawhansett Company, 12,400 spindles, at Putnam; and the Fitchville Manufacturing Company, 13,000, at Bozrah; in Rhode Island, the Green Manufacturing Company, 23,000 spindles, at Phœnix; in 1872, the Manville Mill, of 70,000 spindles, at Manville; the Moss's Manufacturing Company at Westerly, 10,000 spindles; and the Ballou Mill of 50,000 spindles at Woonsocket, in 1873.

In New York, the Lake George Manufacturing Company, 10,000 spindles at Ticonderoga, in 1872.

In Maine, the Barker Mill, at Auburn, 18,000 spindles in 1872, followed by the Lockwood Mill of 32,000 spindles, at Waterville, in 1875.

In Vermont, the North Pownal Manufacturing Company, 16,000 spindles in 1873. In New Jersey, the Millville Manufacturing Company, 30,000 spindles in 1873, and Messrs. R. & H. Adams & Company, 24,000 spindles at Paterson in 1872.

Michigan, the Jonesville Manufacturing Company, 5,000 spindles in 1872.

Illinois, the Chicago Manufacturing Company, 5,000 spindles in 1871, and the Rock Island Manufacturing Company, 5,000 spindles in 1872.

Tennessee, the Brownsville Manufacturing Company, 3,600 spindles in 1874, and the Tennessee Manufacturing Company at Nashville, 14,000 spindles in 1875.

Wisconsin, the Janesville Manufacturing Company of 10,000 spindles in 1874.

North Carolina, the Oakdale Manufacturing Company, 4,000 spindles at Greensboro in 1873, and the Rockfish Manufacturing Company at Fayetteville the same year, 4,500 spindles.

South Carolina shows the Graniteville Manufacturing Company of 23,000 spindles, commenced at an earlier date, the Langley Manufacturing Company at Langley, 10,000 spindles, and the Camperdown Manufacturing Company at Greenville, in 1875, 14,000 spindles.

In Georgia, the Arkwright Manufacturing Company of Savannah,

4,000 spindles in 1873, and some additions of spindles to other previously built mills.

Alabama—the Tallassee Manufacturing Company of 18,000 spindles.

At the present moment efforts are being made for a further extension of manufactures at the South, but the want of capital renders it very difficult to make much progress.

A large mill has been projected, and commenced at Atlanta, and the Eagle and Phœnix Company of Columbus, Ga., are building a new mill, and other projects are under discussion, but at the North the general feeling is that there are spindles enough to supply the present demand for home consumption, and that further increase must depend on the natural growth of the country, and the further development of our export trade, which was severely crippled during the Civil War, and to the revival of which the serious attention of our manufacturers and merchants is now being turned.

CHAPTER XII.

THE number of spindles in the United States in 1874 had reached to 9,415,383, distributed as follows:

TOTAL NUMBER OF COTTON SPINDLES IN THE UNITED STATES, JULY 1, 1874.

Maine	609,898	
New Hampshire	855,189	
Vermont	58,948	
Massachusetts	3,769,692	
Connecticut	908,202	
Rhode Island	1,336,843	
New York	580,917	
New Jersey	150,000	
Pennsylvania	452,064	
Delaware	47,976	
Maryland	110,000	
Ohio	20,000	
Indiana	22,988	
Minnesota	3,400—	8,927,754
Alabama	57,594	
Arkansas	1,256	
Georgia	137,380	
Kentucky	10,500	
Louisiana	15,000	
Mississippi	15,150	
Missouri	18,656	
North Carolina	55,498	
South Carolina	62,872	
Tennessee	47,658	
Texas	10,225	
Virginia	56,490—	487,629
Total number of spindles		9,415,383
Number of mills		847
Number of looms		186,975
Number of spindles, 1874		9,415,383
Number of spindles, 1870		7,114,000
Increase in four years		2,301,383

and the cotton consumed to 1,220,000 bales, or 567,583,873 lbs., which was divided among the following products, viz.: Threads, yarns and twines, 149,000,000 lbs.; sheetings and shirtings, 707,000,000 yards; drills, jeans, flannels, etc., 306,000,000 yards; print cloths, 588,000,000 yards; ginghams, 33,000,000 yards; duck, 30,000,000 yards; bags, 6,000,000.

As print-cloths, as before stated, had been for a few years an article of great demand, it must be inferred that there had been a corresponding increase in printing machinery, although a portion of the cloths were finished as bleached and dyed cambrics, and a large amount consumed for various purposes, such as lining trunks, cheeseboxes, etc., etc., and the following table, taken from the New York "Journal of Commerce," is believed to be correct:

PRINT WORKS AND NUMBER OF PRINTING MACHINES IN THE UNITED STATES, JANUARY 1, 1876.

Woonsocket Co., Providence, R I	12
American, Fall River, Mass	16
Bay State, Fall River, Mass	6
Albion, Coneschocken, Pa	6
Ancona, Gloucester, N. J	10
A. & W. Sprague, Cranston, R. I	30
Cocheco, Dover, N. H	13
Clyde, River Point, R. I	7
Dunnell, Pawtucket, R.I	11
Dundee (Reed & Barry), Passaic, N. J	7
Freeman, North Adams, Mass	7
Garner & Co., Haverstraw, N. J	20
Garner & Co., Wappinger's Falls, N. Y	20
Gloucester, Gloucester, N. J	12
Greenwich, East Greenwich, R. I	7
Hamilton, Lowell, Mass	8
Hunter	8
Hartel, Holmesdale, Pa	6
Hamilton Woolen Co. (Knickerbocker), Southbridge	6
Harvey, Arnold & Co., North Adams, Mass	8
Wm. H. Locke, Passaic, N. J	7
Lodi, Lodi, N. J	3
Manchester, Manchester, N. H	14
Mystic, Medford, Mass	2
Merrimack, Lowell, Mass	18
Oriental, Apponaug, R. I	9
Pacific, Lawrence, Mass	22
Richmond, Providence, R. I	7
Simpson, Philadelphia, Pa	13
Scott (Franklin), Paterson, N. J	7
Saunders, Southbridge, Mass	4
Smith, Philadelphia, Pa	1
William, Bustleton, Pa	4
Total	331

Allowing each machine to produce 200 pieces per day, a fair average, for 40 weeks in the year, this would give a total of 14,400,000 pieces printed annually, of which almost the whole has been used at home, our exports of prints being very small.

Within the last two years a portion of the Fall River production of cloths has been sent to England, where they have probably been printed, and distributed to other markets, with which we have not established an export trade, or with which our commerce was interrupted during the war.

The American calicoes at the Exhibition were very highly commended by the Foreign Judges, and were considered to be fully equal in design, color and execution to those exhibited by any other country, although there was no representation of the higher grades of more expensive goods from France.

The manufacture of ginghams, cottonades, and other cotton fabrics dyed in the yarn, has also been very widely extended, and the goods of the York, Amoskeag, Lancaster, Bates, Everett, Pemberton, Whittenton, Renfrew, and Glasgow Companies, in New England, are well known and appreciated, as are also those of a number of smaller establishments in the Philadelphia district.

Fine lawns and muslins have been made to some extent, although, as has been stated previously, the great bulk of cotton manufactures have been of yarns between No. 14 and No. 40. The manufacture of cotton duck for sails and tents is an American invention, and received great extension during the Civil War, and is widely scattered over the Eastern and Middle States.

Spool cotton is also made in great quantities, the more notable brands being those of the Willimantic Company, at Willimantic, Conn.; the Hadley Company and Merrick Company at Holyoke, Mass.; the Coats Thread Company at Pawtucket, R. I., Green & Daniels of the same place; the Clark Thread Company of Newark, N. J., and Samuel Semple & Sons, of Mount Holly, Burlington County, N. J.; the Coats and Clark companies being originally offshoots from the parent establishments in Scotland.

Cotton bags, woven without seam in the loom, are also of American introduction, and are made from the waste left in the manufacture of finer fabrics, mixed with stained or lower grades of cotton.

Small wares, such as tapes, braids, lamp-wicking, suspender webbing, etc., etc., are extensively made, principally in Massachusetts, Rhode Island, and in and about Philadelphia.

The accompanying table shows the statistics in a condensed form of several of the larger towns and cities, which owe their prosperity, and in most cases their origin and existence, to the cotton manufacture, having been first established in situations where a natural waterfall rendered a great amount of power available; and in addition to these, the whole area of the New England States is studded with cotton mills, some of them of great size and importance, wherever suitable water power was to be found, in many cases compensating for the droughts of summer by the addition of steam.

STATISTICS OF LOWELL, 1876.

NAME OF ESTABLISHMENT.	Date of Incorporation.	No. of Spindles.	No. of Operatives.	Lbs. Cotton consumed annually.	Yards Cloth produced annually.
Merrimack Mfg. Co.	1823	158,464	2,700	6,344,000	37,700,000
Hamilton "	1825	56,080	1,225	3,900,000	14,040,000
Appleton "	1828	42,488	600	4,992,000	12,480,000
Tremont & Suffolk Mfg. Co.	1830	93,528	1,400	7,280,000	19,760,000
Lawrence "	1831	92,000	1,750	9,100,000	22,100,000*
Boott Cotton Mills.	1835	112,752	1,875	6,760,000	23,920,000
Massachusetts Cotton Mills.	1839	101,720	1,475	9,256,000	27,768,000

STATISTICS OF LAWRENCE, MASS.

Atlantic Cotton Mills.	1846	86,880	1,000	8,800,000	24,500,000
Pacific Mills (also worsted).	1852	135,000 worsted 25,000	5,000	6,000,000	42,000,000
Pemberton Mills (a'so woolens).	1860	28,000	675	1,710,000	3,000,000
Washington " " "	1858	20,000	300	675,000	warps, 5,200,000
Everett "	1860	33,280	775	2,250,000	7,000,000
Lawrence Duck Co.	1853	7,500	225	1,500,000	1,700,000

STATISTICS OF MANCHESTER, N. H., 1876.

Amoskeag Mfg. Co.	1831	135,000	4,000	13,000,000	36,400,000
Stark Mills.	1838	45,000	1,200	6,760,000	12,376,000
Manchester Mills (also worsted), reorganized.	1873	75,000	2,940	4,160,000	28,600,000
Langdon Mfg. Co.	1857	33,036	500	1,560,000	4,940,000

STATISTICS OF LEWISTON, MAINE, 1876.

Lincoln Mill.	1846	21,746	407	1,040,000	3,000,000
Bates Mfg. Co.	1850	56,196	1,250	2,023,114	8,151,000
Hill "	1850	51,000	1,000	2,500,000	8,000,000
Continental Mills.	1866	70,000	1,200	6,000,000	15,000,000
Lewiston " (also jute).	1853	25,000	850	2,800,000 jute, 1,011,000	....
Androscoggin Mills.	1860	58,450	1,100	4,350,000	7,000,000 bags, 2,000,000
Barker Mill (Auburn).	1870	18,576	250	855,000	2,000,000

* Lawrence Company also, 780,000 dozen hosiery and 46,800 shirts and drawers.

Statistics of Fall River, 1876.

Name of Establishment.	Date of Incorporation.	No. of Spindles.	No. of Operatives.	Lbs. Cotton consumed annually.	Yards Cloth produced annually.
American Linen Co	1852	82,512	1,000	3,825,000	21,000,000
Annawam Manufactory	1815	10,116	140	450,000	2,150,000
Barnard Mfg. Co.	1874	28,400	340	1,575,000	9,000,000
Border City Mills	1872	72,144	900	3,712,500	20,500,000
Chase "	1871	43,480	425	2,025,000	12,000,000
Crescent "	1871	33,280	340	1,462,500	5,750,000
Davoe "	1867	30,496	375	1,575,000	5,000,000
Durfee "	1866	87,424	950	4,275,000	23,000,000
Fall River Manufactory	1813	25,902	330	1,350,000	7,000,000
" " Print Works	1848	13,600	175	607,500	3,500,000
Flint Mills	1872	45,360	450	2,137,500	12,500,000
Granite Mills	1863	76,920	900	4,050,000	21,500,000
King Philip Mills	1871	37,440	425	1,350,000	5,500,000
Mechanics' "	1868	53,712	550	2,587,500	14,000,000
Merchants' Mfg. Co	1867	85,570	800	4,162,500	22,500,000
Metacomet Mills	1847	23,840	325	1,125,000	6,500,000
Montauk "	1871	7,200	125	1,125,000	2,000,000
Mount Hope "	1867	9,024	135	303,750	1,225,000
Narragansett "	1871	27,920	325	1,462,500	8,250,000
Osborn "	1871	37,232	425	1,912,500	11,000,000
Pocasset Mfg. Co	1822	36,744	510	1,417,500	7,500,000
Richard Borden Mfg. Co	1871	42,528	450	2,025,000	12,000,000
Robeson Mills	1867	21,632	275	1,125,000	6,500,000
Sagamore "	1872	37,672	425	1,800,000	10,500,000
Shove "	1872	37,504	425	1,912,500	11,500,000
Slade "	1871	37,040	350	1,800,000	10,000,000
Stafford "	1871	34,928	350	1,800,000	10,000,000
Tecumseh "	1866	42,156	400	2,025,000	12,000,000
Troy Cotton & Wool Manufactory	1814	38,928	400	1,800,000	10,250,000
Union Mill Co	1859	44,784	475	2,250,000	12,000,000
Wampanoag Mills	1871	27,920	325	1,462,500	8,250,000
Weetamoe "	1871	34,080	350	1,800,000	10,000,000
Fall River Merino Co	1875	1,560	60	337,500	9,000,000
Total		1,269,048	14,270	62,628,750	343,375,000

The question of the comparative economy of steam and water-power has often been discussed, and results drawn which have proved erroneous, for want of correct data in the premises.

As before stated, many cotton mills were started in the seaboard towns of New England, in the belief that steam was as cheap a motor as water, and nearly if not all of those east of Fall River have proved unsuccessful as investments.

The writer had occasion to make a careful examination of the cost of power at some of these mills a few years since, and arrived at the following results:

	Mill No. 1, Newburyp't.	Mill No. 2, Rockport.	Mill No. 3, Newburyp't.	Mill No. 4, Fall River.
Number of spindles	17,040	17,904	26,976	34,848
Horse-power required	275.75	291	327.5	450
Tons of coal per annum	1,612	1,873.5	2,213	2,632
Cost of " "	$10,823.24	$12,486.78	$14,560.00	$18,766
Engineer, fireman, oil, etc., including delivery of coal	2,074.28	3,556.92	2,701.29	2,500
Total cost	$12,987.52	$16,053.70	$17,261.29	$21,266
Cost per horse-power per annum	47.10	55.12	52.70	45.27

Including the Fall River Mill, which agrees with two other mills there very closely, the average cost of the 4 mills for fuel, oil, and labor per annum per horse power is $50.04, or for the mills east of Boston, where coal was higher, $51.64—to which must be added about $20 per horse power for interest and depreciation on plant, or 20 per cent. of a fair average cost of $100 per horse power for engine, boilers, and setting, engine house and chimney; of which the engine would cost $\frac{1}{4}$ at 10 per cent. depreciation; boilers, $\frac{1}{4}$ at 20 per cent.; and buildings and chimney $\frac{1}{2}$ at 6 per cent., making an average of about 12 per cent., which, with interest, etc., would bring the whole allowance up to about 20 per cent. This would give a total cost for steam power of $70 per horse power per annum, which may be assumed as the average cost in quantities of from 2 to 500 horse power near the sea coast of New England. This would, of course, be increased or lessened in different localities by the cost of coal. Mr. George H. Corliss, of Providence, R. I., the eminent steam-engine builder, estimates the cost with his improved engines, including 20 per cent. as above, to be 16.22 cents per day per horse power, which, for 300 working days, would give $48.66 per annum.

The above figures were, however, taken from the actual running accounts of mills, in ordinary years. Now at Lowell and Lawrence, the annual water rent per horse power paid to the Water Power Companies is $5.00 per annum; and if the cost of wheels, pits, and flumes be estimated at $100 per horse power, there is to be added $7.00 per annum for interest, and $5.00 per annum for depreciation, making in all, for power, $17.00 per annum. To this should be added the further sum of $3.00 per annum, for heating and dressing, giving a total expenditure of $20.00 per annum per horse power of water, as against $70.00 for steam.

At the Androscoggin Mills, Lewiston, Me., the cost of water power and heating is made up as being $14.10 per horse power per annum; that of steam, as being $67.92; interest and repairs not being included in either case. In this account the coal was taken at $8.50 per ton. There are many mills in the New England States, where the whole cost of water power, including dam, wheels and canals, has not been over $100 per horse power; and allowing $12.00 per annum for interest and depreciation, and $3.00 more for heating and dressing, the cost in these cases would be only $15.00 per horse power yearly.

Against this positive difference in the cost of power, we must set off the less cost of freight in such localities as Fall River, which, however, owes much of its success to the general system of management pursued there, and to various causes which are not within the scope of this article, and the discussion of which would occupy much time and space, and involve opposing arguments.

Steam power in smaller quantities, say less than 100 horse power, would cost more than the above rates, while water power would usually decrease in cost, from the need of less expensive dams and wheels.

The average cost of steam engines may be taken as being:

For 800 to 1,000 horse power	$20 per horse power
For 500 to 600 "	30 " " "
For 200 to 300 "	40 " " "
For 50 to 100 "	50 " " "
Boilers and setting,	30 to 40 " "
Engine and boiler house, chimney, etc.	30 " "

"The Engineer" gives the cost in England for 100 horse-power engine, with boilers and "plant" complete, as being $75 per horse power, and the cost of fuel, etc., per annum, at $50 per horse power. The cost of a first-class turbine wheel is given by the same authority as $3,500 for 100 horse-power, or $35 per horse-power.

The turbine has almost entirely taken the place of the cumbrous over-shot or breast wheels, and may be procured at very reasonable prices.

While the Fourneyron wheel, as improved by Mr. Boyden, has been generally used in the large manufacturing establishments of New England, the attention of engineers and inventors has been turned to the production of a less expensive wheel, of equal efficiency, and there are now a number in the market giving excellent results, which are cast in one piece, instead of being "built up" with bronze or other sheet metal buckets riveted into cast-iron plates or flanges. Several of these were tested at the Centennial Exhibition, and

The Risdon wheel, which gave over.............. .... 87 per cent.
The National wheel   "   "   "   83   "
The Geyelin wheel   "   "   "   83   "
The Tait or Centennial wheel which gave over.................. 82   "
The Tyler wheel which gave over......... 81   "
and the Hunt wheel which gave over........................ 80   "

are all well made and reliable wheels.

Besides these there were several others giving over 75 per cent., which may be considered as the *maximum* effect of the old style of wheels.

Of these wheels, the Geyelin was of the Jonval pattern, with a direct downward discharge of the water; the others were all inward and downward, as is also the case with the Swain wheel, which was not on exhibition, but which is very largely used in New England, and the form of bucket of which is the progenitor of those of several of the above-named wheels.

Mr. James B. Francis, of Lowell, has obtained with this wheel a result of over 83 per cent. of effect, but the wheel is like the Boyden, a "built" and expensive one, compared with those mentioned.

This method of central discharge of the water is believed to be of American origin, the type of this class having been the "Howd wheel," patented by Samuel B. Howd, of Geneva, N. Y., July 26, 1838.

Water wheels can not be strictly considered as parts of "Cotton Machinery," but their common use and economy as motors in America seem to excuse the above digression.

I have aimed to trace, as concisely as possible, the growth of the cotton manufacture in the United States up to the present time, and the dates of the more important American inventions which have contributed to its success, and have endeavored to omit nothing which could be condensed within the space I have allowed myself for this memoir; and I cannot bring it to a close better than by the insertion of the following article from the New York "Herald," which I am permitted to use by the kindness of the author, Edward Atkinson, Esq., of Boston, which states clearly our present position, as a Nation, in regard to the production of the raw material; and also the admirable analysis of the cost of manufacture in 1838 and 1870, prepared for me by Mr. William A. Burke, of Lowell, the treasurer of the Lowell Machine Shop, and read by him before the New England Cotton Manufacturers' Association, which is inserted in the Appendix:

EXTRACT FROM LETTER OF EDWARD ATKINSON TO THE "NEW YORK HERALD."

"The commanding position of the United States in respect to the production of cotton has long been admitted, but it seems probable that few even of the manufacturers themselves have been fully aware of the strong position in which the cotton manufactures of the United States now stand in relation to other countries.

"A HISTORY OF COTTON.

"The subject of the production of cotton opens so wide a field that it is hard to know where to begin or end. There is no other product that has had so potent and malign an influence in the past upon the history and institutions of the land, and perhaps no other on which its future material welfare may more depend. Cotton belongs to this continent. When the Spaniards first entered Mexico the natives were found to be clothed in cotton, and the art of weaving and dyeing had been carried to a high state of perfection for that time among them. Then, as now, the best and most prolific varieties of the cotton plant existed there, and the plant is doubtless indigenous in Mexico. In the United States, a century ago, cotton was scarcely known as an important production, and not until the invention of the saw gin, by Eli Whitney, in 1792, did it become so; that invention renewed the life of slavery. To-day the United States furnish all the cotton used in their own limits and in Canada, and nearly three fourths the quantity consumed in their own limits and in Europe combined. There are no data by which the quantity produced and consumed elsewhere can be determined accurately, the production of Asia and Africa being unknown; but the inhabitants of these two continents are clothed in cotton to a very large extent of their own production and manufacture.

"COTTON MANUFACTURE.

"In respect to the cotton manufacture the world may be divided into two sections—that which still adheres to the hand work, and which is by far the largest and most populous section, and that which uses complex machinery worked by water or steam power. It is common to name the divisions "civilized" and "uncivilized"; but, if there had been no previous reason for hesitating to apply these terms, the wonderful exhibition from China and Japan at Philadelphia might well teach us a lesson in modesty.

"Among the machine-using nations it may, perhaps, be rightly claimed that the United States takes the lead; not that we can assert superiority in all, or perhaps in any, special machines, but that our people adopt machinery more quickly than others, and adapt it to a greater variety of purposes. The object of this paper is to mark the progress we have made in the cultivation of cotton and in the application of machinery to its manufacture, and also to forecast the work we may have yet to do.

"COMPARATIVE IMPORTANCE TO FABRICS.

"Among the three common fibers—wool, flax and cotton—which constitute the principal materials for the clothing of the human race, cotton is the most important, because it is ready for treatment by machinery as soon as it is gathered; because its conversion into cloth is least costly, and because its use for clothing is most conducive to health in respect to the largest portion of the population of the world. It is a non-conductor of heat and of electricity, while flax is the reverse; it is easy to spin because Nature begins to twist it in the boll, and each fiber is like a twisted ribbon, a little thicker at the edges than in the middle; hence the fibers interlock and adhere to each other to their very points. The great inventions in cotton spinning have not been in the twisting, which is a comparatively simple matter whether compassed upon one spindle or many, but in the extension of the strand both before and after the twisting begins.

"PROCESSES OF PREPARATION.

"The processes applied to the fiber in order to convert the bale of cotton into yarn for weaving are of three kinds—first, to clean and straighten the fibers and lay them alongside each other in a thick and heavy strand; second, to extend that strand with a constant doubling of two or more ends into one in order to get the strand even; and third, to combine the further extension and doubling of the strands with the twisting. The extreme accuracy required in the working of the machinery will be best appreciated from the fact that the No. 14 yarn, of which the coarse standard sheeting is made, weighs sixty hundredths of a grain to one yard, while the yarn in a common lawn of which a woman's summer·dress is made, No. 70, weighs twelve hundredths of a grain. It follows that all the complex machinery and the twelve to fifteen processes through which the cotton must pass from the bale to the spindle are worked within the limit of about half a grain in the result, the two numbers named representing substantially the whole cotton spinning of the United States. The number indicates the number of skeins or hanks of 840 yards each in one pound avoirdupois, or 7,000 troy grains, the cotton spinners' tables being based on troy grains and avoirdupois ounces and pounds.

"If we consider our production of cotton in the light of a service rendered, we then find that it stands first in rank among the material services which we render to humanity. In the cotton factories of Europe and the United States there are a little over 68,000,000 spindles, worked by about 1,000,000 men, women and children. In the operation of these spindles a little more than 6,000,000 bales of cotton, of the average weight of American bales, are annually converted into 10,000,000,000 yards of cloth, averaging one yard wide and four yards to the pound, or 10 pounds to a piece of forty yards, or into the equivalent of such cloths in other fabrics. As nearly as the writer can ascertain, the fabric called by the trade a four-yard sheeting is about the average fabric made on the cotton spindles of the world. In this country the average would be heavier; in Great Britain lighter. The fabric made by the Lawrence Manufacturing Company, known as LL, is a representative of this average.

"This quantity of cloth would furnish 500,000,000 persons five pounds or twenty yards each annually. Of the 6,000,000 bales of cotton the United States now furnishes about 4,500,000 in each year, and our proportion is year by year increasing. The eight last crops, raised by the labor of freemen, exceed the eight last crops before our civil war, then raised mainly by the labor of slaves, in the number of more than 1,500,000 bales. If, then, it is a service to men to provide for them the largest quantity of the material that best meets their need for clothing, in this one respect our rank is assured.

"AMERICAN CAPABILITIES.

"Then let us mark the extent to which we have yet trenched upon our resources. In this production less than 2 per cent. of the area of the cotton States is now used. What we may yet accomplish may be better comprehended by considering the condition of a single State. We will select Texas as being the State now making the most rapid progress in population, production and wealth. Few persons can realize the facts in regard to this great State except by comparison. In area it exceeds the German Empire by about 60,000 square miles; it has land and climate fitted for the growing of almost all the products of the temperate zone; it is underlain to a large extent with coal. But, in respect to cotton, on less than one half of one per cent. of its area it last year produced one half of all the cotton consumed in the United States, and 4 per cent. of its area would be capable of producing all the cotton now consumed in Europe and the United States, or 6,000,000 bales. Whenever the fertile land of Texas, which constitutes nearly three fourths its area, is settled with the same density of population as Massachusetts, one person to each three acres, it will contain nearly 40,000,000 people.

"Under what conditions is this work of cotton production now accomplished or yet to be done? No longer by the forced labor of the slave upon the plantation, but by the labor of freemen and largely of freeholders on the farm. In most of the States where it is now grown, cotton constitutes the salable or money crop of the farmer, who, in other respects, is becoming entirely independent as to his subsistence. Raising food and meat to a greater extent than ever before, the Southern farmer still finds in cotton the means wherewith to furnish himself with money for other purchases. Cotton being therefore more and more the surplus crop or profit of the farmer as distinguished from the planter of old time, it becomes more difficult to determine its cost, its annual quantity until each year's crop has been actually delivered, or the price at which its production will be checked. In Texas, the State that has increased its crop about 80 per cent. over the largest ante-war crop, by far the largest portion is now raised by farmers owning their own lands. Her last crop was nearly 700,000 bales, and within one or two years at farthest it will be 1,000,000, mostly cultivated by white labor.

"COST OF PRODUCTION.

"In answer to a very extended inquiry lately made, the writer has received estimates of the cost of the production of cotton ranging from six to fifteen cents per pound, the latter cost, however, having been given by one who on 600 acres of land made only four bales the previous year. The general range of the estimates of cost were six to ten cents. But one answer to the question of cost was the most significant. One said: "I have a nephew twenty years of age who, without the least detriment to his schooling, and working Saturdays, produced 4 bales of cotton." It may be asked what did this lad's cotton cost to produce? The average estimate of cost is nine and six tenths cents per pound; those who give the higher rates basing their estimates upon the purchase of provisions at present prices; those who give the estimate of six to seven cents basing them upon provisions being raised on the same farm. The significant fact in all the estimates is that the lowest come from Texas, North Carolina and Georgia, which are essentially farming States, while the highest come from Mississippi and Louisiana, the States which were formerly *par excellence* the country of the large planters.

"IMPROVEMENTS.

"According to these returns, the centennial year is also marked by greater improvements than ever before in the selection of seed, in the improvement of tools, in the use of fertilizers, and in the average crop per acre, positive evidence having been given of the production of 2,500 pounds of lint or clean cotton on a single measured acre in Georgia. It was not claimed that this had been or could be profitable, but it is significant of the experiments that are being tried in many places. The average estimates of profitable work range from 400 to 1,000 pounds of lint or clean cotton per acre, according to the quality of the soil and the kind of work done or the fertilizers used. The last ten years have also witnessed the conversion of the seed of the cotton plant into many useful articles but little known before.

" In respect to the estimates of the cost of raising cotton it does not yet appear that any very accurate data exist under the new system ; skillful men, who, immediately after the war, feared utter ruin unless the price could be maintained at 20 cents per pound, now admit having made a fair profit at 10 cents. It may be doubted whether the cost can ever be defined. If the farmer can raise an ample supply of grain, vegetables, meat and fruit for subsistence, and can also produce more or less cotton for sale, the cotton represents profit or surplus rather than cost, and under such circumstances its production would not cease, although it might be checked, even if it should decline to 6 or 8 cents per pound.

"OLD AND NEW METHODS.

" The future increased production of cotton in the United States and the time within which our staple will take the place of all inferior grades is, therefore, only a question of numbers and intelligence. In respect to intelligence it is not to be questioned that the planter of old time had far more skill than many of the farmers of the present day, but the system of labor to which that skill was applied imposed bad conditions that could not be surmounted, and it enforced the use of tools and methods unfit for the purpose. These methods may have assured prosperity to the few at the cost of the many, but it was the high price and not the low price of cotton that limited the extension of the crop. Twenty years since every bale that could be made by the force then upon the cotton field was required for use, and under the steadily advancing price the capital needed for opening new fields as steadily increased with the advancing price of slaves, until, in 1860, it cost 50 per cent. more to buy and stock a cotton plantation to raise the cotton for a given factory than it did to build the mill and fill it with machinery. All this has changed, and in the five years last past more than a million persons have migrated from other States or from abroad to the fertile lands of Texas, and the independent freeholder will only be prevented from making more and more cotton each year by the low price and not by the high price that it may bring. That no such check is very near may presently be made apparent.

"IMPROVEMENTS NEEDED.

"In one respect great improvement is needed, and but little has yet been made. The separation of the lint from the seed is the process that should be mostly fitly accomplished, but which is now most rudely done. The best saw gin of the usual construction, unless most carefully attended, tears, breaks, doubles and otherwise injures the staple, and but a small proportion of the cotton now made is delivered to the spinner in the best condition. Two new cotton gins were exhibited at Philadelphia, which promise excellent results—the roller gin made by Messrs. Platt Brothers & Co., of England, and the needle-point gin made by the Messrs. Remington, of Ilion, N. Y. If these machines can be made to produce quantity in ratio to the quality of the staple which they deliver, their wide introduction cannot be long delayed.

"PACKING, ETC.

"The methods of packing, covering and handling cotton in the United States is now unfit in the extreme, and as the competition becomes greater with declining prices it is to be hoped and expected that better methods will be adopted. At present it is alleged that it is not profitable to attempt better methods, but the time can not be far distant when the bale of cotton will be as carefully prepared and protected as the bale of cotton fabrics. Such care and attention was formerly impossible. It must be remembered that the slave system repelled and degraded free labor; its malign effect was never more tersely expressed than by Henry A. Wise, of Virginia, who said, "The negroes skin the land and the white men skin the negroes." But all this has passed by, and the professor of a South Carolina college who was sent out of the State because he said, when the ordinance of secession was presented to him, "That is not what South Carolina wants—she needs manure," is now engaged in providing that valuable commodity, being at the head of one of the large works established since the war for converting the phosphatic rocks that underlie her coast lands into the material that her exhausted soil so much required.

"THE GOOD TIME COMING.

"With local self-government assured and the colored race fully protected by the urgent competition for their labor, it cannot now be doubted that the fertile soil and the mild climate of the South will soon attract to other portions as large an immigration as is now pouring into Texas; and as new branches of industry are established and a more dense population grows up or comes in, even though the negro should quit the fields and take to other occupations in towns and villages, as it is alleged he will do, there can be no doubt of the continued increase of the cotton crop."

EXHIBIT OF MACHINERY AND FABRICS AT CENTENNIAL.

SUPPLEMENTARY CHAPTER.

THE cotton manufacture of the United States, as illustrated at the Centennial Exhibition, although very complete in its collection of the various fabrics made in the country, was very deficient in its illustration of the processes by which they were produced, many of the best machines being entirely unrepresented, and some of those shown not being in operation, while in no case was a complete system of cotton machinery shown by any exhibitor.

There were, however, machines from various sources on exhibition, from which we can trace the operation of the various processes which the cotton undergoes in its progress from the bale to the cloth, and we will endeavor to note these in their proper order, with some comments on their peculiarities.

First in order was the opening and picking machinery of the Kitson Machine Co., of Lowell, consisting of two machines—the first one for opening the cotton, from the bale, and partially cleaning it from seed and leaf, and forming it into a lap, which is taken to the second machine, where it is still further cleaned, and where three of these laps are united into the one which goes to the card.

The second machine did not differ essentially from the "Lappers" in ordinary use in the United States and England, but the first one had a radical modification, in the addition of a toothed cylinder (or pair of disks, both forms being employed) to tear open the hard masses of cotton as they are taken from the bale, and to loosen them, before subjecting them to the blows of the "beater," whose office it is to knock out the seeds and sand, while the cotton is carried forward by a current of air against a wire gauze cylinder, which forms it by its revolution into a sheet or lap. The processes of opening cotton have varied considerably in times past, according to the character of the staple to be treated; but the mode most commonly adopted of late years, for the great bulk of middling and ordinary cottons, has been that of the "Beater" so-called, a straight steel bar, of which two or three were set on arms, parallel with and 7 to 12 inches distant from a shaft revolving from 1,200 to 1,800 times per minute, thus giving the beaters a velocity of from 5,000 to 8,000 feet per minute. These bars strike the cotton as it issues slowly from a pair of feed rollers, and by their velocity drive the seeds and heavy substances downward through a grating underneath, while the cotton is carried on with the beaters till it is sucked off by an air draught on the side opposite to the feed rolls —as above mentioned.

With loosely packed and dry cotton this mode is not objectionable, but, when it has been heavily compressed and subjected to moisture, the hard "mats" thus formed are very difficult to open, causing great strain and wear on the machines, and consuming an unnecessary amount of power, besides causing more or less injury to the staple, and it is to obviate these difficulties that Mr. Kitson's "Atmospheric Opener" was constructed. These machines are now being introduced to great satisfaction in many large cotton mills—both in America and England.

The next machine in order is the card—of which there was not one of the ordinary form on exhibition, the only machine shown being Messrs. Foss & Percy's "underflat" card, which aims to do double the work of the ordinary card, with only $\frac{1}{3}$ more power, and only the same amount of floor space in the mill.

The novelty of this machine lies in the arrangement by which the "Card Flats" are held up by springs to the under side of the carding cylinder, thus giving double the working surface; and in so revolving the cylinder that it strikes the cotton downward instead of upward from the feed rolls, thus knocking any dirt and sand which may have escaped the picker down into a space prepared to receive it, instead of carrying it up and lodging it in the "Top Flats." This machine is but of very recent introduction, and it is not yet possible to speak authoritatively of its merits, but it was favorably esteemed by the judges, and is now being thoroughly tested, with excellent results, in several of the mills at Lowell and other manufacturing towns.

Of the next process, the "Drawing-Frame," there was only a single specimen exhibited, and that one not in operation: from the Saco Water Power Machine Shop, of Biddeford, Me.

This machine was thoroughly well built and finished, and may be considered as a fair specimen of the machines at present used for that purpose.

The same establishment also sent a very perfect set of Roving or "Fly Frames," three in number, of excellent finish and construction, and a self-acting mule, of the Parr-Curtis pattern, which may all be taken as good specimens of the best type of cotton machinery in use at the present day.

The Providence Machine Company also exhibited an excellent set of roving frames in operation, which were noticeable for the high velocity and great ease of their movements, but which differ little in essential points of construction from those of the Saco Water Power Machine Company.

Messrs. Fales & Jenks, of Pawtucket, R. I., had on exhibition the "Combined Fly Frame and Speeder" of Messrs. Mayer & Chatterton, which had several excellent features, combining very ingeniously some of the best points of two systems ; having the long flyer of the speeder separated from the spindle, thus admitting of economy of time in dressing and the more perfect "wind motions" of the fly frame.

No examples of the spinning frames which have been in ordinary use for many years were on exhibition, but two varieties of the new light ring spindle, which is rapidly being introduced, were shown in operation, viz., one frame of the "Sawyer spindle," by Messrs. Geo. Draper & Sons, of Hopedale, Mass., and one frame built by the Lowell Machine Shop, and exhibited by the Willimantic Co., in which half the spindles were of the Sawyer pattern, and half of a combination of the "Sawyer" and "Pearl" patents. This frame was shown running at a very high speed, and spinning No. 160 yarn ; but, as both these spindles have been described in another place more fully, it is unnecessary to say more about them here.

Messrs. Draper & Sons also exhibited a twister with the Sawyer spindle, a spooler with the "Wade" bobbin holder, a new and very ingenious improvement, and a Warper, containing several new and valuable inventions.

The Lewiston Machine Shop, of Lewiston, Me., also contributed a Warper, of the very best plan and construction, and one which is in very extensive use and highly commended in many of our best mills.

The display of looms was quite large, and contained several patterns of the well-known Crompton loom, from Worcester, Mass.; the Knowles loom, from the same place ; the Thomas loom, from the Lewiston Machine Shop ; the Lyell loom, which was shown weaving jute canvas, 8 yards wide, grain-bags, sheetings, and women's corsets, the latter, by an application of the Jacquard motion to the warp, being produced shaped to the body, and the loom weaving 8 at once.

Other looms were exhibited by Thomas Wood, of Philadelphia, Jas. Long, Bros. & Co., Philadelphia, L. E. Ross, Providence, R. I., Wolfenden, Shove & Co., Cardington, Pa. All these looms were excellent for their intended purposes, and very thoroughly illustrated that branch of manufacture, especially so far as related to figured or "fancy woven" fabrics.

Messrs. George Draper & Sons also exhibited loom-temples such as are in universal use in the United States, and double adjustable spinning rings.

The Willimantic Co. of Willimantic, Conn., made a very fine display of cotton in its various states of progress from the bale to the yarn, and showed the two very ingenious automatic machines invented for them by Hezekiah Conant, for spooling sewing cotton, and for ticketing the spools.

Messrs. Butterworth & Co., of Philadelphia, exhibited drying cans, for bleached or printed fabrics; Messrs. Palmer & Kendall, of Middletown, Conn., have a very ingenious drying and tentering machine for mosquito nettings, lawns, and other light cloths; and Messrs. Hope & Co., of Providence, R. I., showed two patterns of their very ingenious Pantograph engraving machines.

There were some smaller exhibits, but the above list comprised all the machinery of any importance in the American department, while the display from other countries was so meager as to allow of no particular comparison.

The display of fabrics was much more extensive, consisting of 142 exhibits, ranging, with one or two exceptions, from No. 8 to No. 40 yarn, as the basis, and comprising a very great variety of styles of cloth, from cotton duck to fine muslins, ginghams, and printed calicoes, from different parts of the Union.

A full list would be simply a repetition of the official catalogue, and it may be condensed by saying that Maine was represented by the following exhibits, viz: Cotton duck from the Westbrook Manufacturing Company of Portland; tickings, cottonades, cheviot shirtings, seamless bags, sheetings, shirtings, jeans, quilts, ginghams, and silesias from the different mills at Lewiston; cottonades, denims, tickings, dress goods, skirtings, and shirtings from the York mills at Saco; sheetings, shirtings, drills, and jeans from Biddeford; and sheetings and shirtings from the Cabot mills at Brunswick and the Farwell mills at Lisbon. New Hampshire also made an extensive display, consisting of tickings, denims, awning stripes, cotton flannels, ginghams; fancy shirtings, jeans, drills, duck, seamless bags, printed calicoes, sheetings, and shirtings, from the mills at Manchester; quilts and sheetings from the Monadnock mills, at Claremont; cotton flannels, sheetings, shirtings, and print cloths from Nashua, sheetings and shirtings from Great Falls, and knitting cotton from Morse, Kaley & Co., of Milford.

Massachusetts was largely represented by print-cloths, printed calicoes, shirtings, and muslins, from Fall River; sheetings, shirtings, drillings, cotton flannels, printed calicoes and furniture-coverings, and bleached and dyed cambrics from Lowell; printed calicoes and lawns, cottonades, tickings, cheviots, denims, and dress goods from Lawrence; spool cotton, lawns, muslins, thread, twine, and yarns, from Holyoke; cotton flannels from Chicopee, denims and dress goods, from Palmer; sheetings and sateen jeans from Salem; sheetings and shirtings from New Bedford and Waltham; ginghams and yarns from South Hadley; ginghams, skirtings, and dress goods from South Adams; cottonades, shirtings, tickings, denims, awning-stripes, and dress goods from Whittenton; ginghams from Lancaster; and print cloths, cambrics, and printed calicoes from Southbridge. Rhode Island, as might be expected, as the birth-place of the cotton manufacture in America, made a wide and varied exhibit of brown and bleached cottons and dyed and printed fabrics, from a number of mills in different parts of the State, all having their business headquarters at Providence. Woonsocket, Pawtucket, Warren, Westerly, Lonsdale, Pontiac, and other manufacturing towns were well represented; and the lawns and muslins of the Lonsdale Co., and the similar fabrics from the Berkeley Co., are worthy of especial notice from the perfection of their manufacture and the fineness of the fabrics; the sateens of the Berkeley Co. being made of No. 100 warp and No. 150 weft, and supposed to be the finest goods made in the United States. Spool cotton of excellent quality was also exhibited, and a great variety of bleached and colored goods from different bleacheries and print-works.

Connecticut sent denims, ducks, cheviots, tickings, and fancy stripes from Norwich; shirtings and cambrics from Wauregan, Putnam, and Taftville; mosquito nettings from Middletown; and sewing cotton from Willimantic, the latter being of especial merit for its smoothness, softness, and strength, and is mentioned in another place, in connection with the machinery employed to produce it.

New York had fewer exhibitors, sheetings and shirtings from Utica and the "New York Mills," silesias, cambrics, etc., from Saratoga, comprising her list.

New Jersey was more fully represented, and sent spool cotton from Newark and Mount Holly; ginghams, print-cloths, and printed and dyed fabrics, from Gloucester; cotton towelings and dusters from Paterson; bleached shirtings, cambrics and silesias, printed linings, umbrella cloths, window hollands, tilloting cloths, etc., from Millville, and cotton yarns and wicking from Wortendyke.

Pennsylvania showed awning stripes, tickings, printed and dyed calicoes, and muslins in great variety—cottonades, ginghams, cotton yarns, and a great variety of cotton fabrics of medium fineness, from the immediate vicinity of Philadelphia; cotton flannels, nankeen, and tickings from Lancaster; counterpanes and quilts from Allentown and Lancaster; and tickings from Linwood.

Delaware sent tickings from Wilmington, and warp yarns from New Castle.

Maryland took the lead in cotton duck, with three very fine exhibits from Baltimore.

North Carolina sent sheetings, from Great Falls, Rockingham Co.

Georgia exhibited very good sheetings from the Alabama and Georgia Manufacturing Company, of West Point.

Mississippi contributed excellent heavy cotton, and mixed cotton and wool fabrics, from the Mississippi Manufacturing Company of Wesson, and also excellent yarn from F. E. Whitfield, of Corinth, which had been manufactured directly from seed cotton, by an apparatus attached to the card, which performed the same purpose as the gin.

Indiana completed the list of the States of the Union represented, with staple heavy sheetings from Evansville.

Canada exhibited staple sheetings, shirtings and yarns, tickings and bags, from Hamilton, Dundas, and Toronto, and New Brunswick sent yarns from St. Johns; and these should fairly be included in the list of distinctively *American* exhibits, as the skilled labor which they have drawn from the United States has been a great element in their success.

In taking a general survey of the subject, a tendency to the manufacture of finer fabrics was noticeable in Massachusetts, and especially in Rhode Island, and also in the newer mills in Maine. New Hampshire excelled in the heavier goods, such as denims, drills and tickings, though the finest tickings are from Pennsylvania, and Maryland made a specialty of duck; Massachusetts and Rhode Island took the lead in ginghams, print cloths, and dyeing and printing, in the quantity produced, although for excellence in dyeing and printing Pennsylvania and New Hampshire fairly disputed the palm with them; while for great variety of minor manufactures, the region directly about Philadelphia excelled.

The Southern States made a small but creditable exhibit, and the time will come when with peace, and attention to industry instead of politics, they will make the greater part of their ordinary clothing fabrics, while the North, as it is now doing, turns its attention to finer goods. With this short comment we will close this memoir of the growth of the cotton manufactures of the United States up to the period of the Centennial Exhibition of 1876.

95

APPENDIX A.

PAPER READ BY WILLIAM A. BURKE, ESQ., OF LOWELL, BEFORE THE NEW ENGLAND ASSOCIATION OF COTTON MANUFACTURERS, OCTOBER 25, 1876.—STATISTICS RELATING TO THE COST OF MANUFACTURING DRILLINGS AND STANDARD SHEETINGS IN 1838 AND 1876.

"MR. PRESIDENT AND GENTLEMEN : I have been requested by your Board of Government to present to this meeting some statistics of the cost of manufacturing drillings and standard sheetings in the years 1838 and 1876.

"These statistics were collected at the request of Mr. Samuel Webber, to be used in his work, soon to be published, on the 'Centennial History of the Cotton Manufacture' in this country.

"The mills I have chosen are the No. 1 Mill of the Boott Cotton Mills, in Lowell, Mass., and the mill of the Jackson Company, in Nashua, N. H.

"The year 1838 is as far back as the records were complete enough to be relied on.

"For convenience of comparison, the items are arranged in a column for each of the years 1838 and 1876.

"The No. 1 Boott Cotton Mill was one of four mills built and equipped ready for operation by the 'Proprietors of Locks and Canals on Merrimack River,' and was started in 1836. It was filled with machinery for making drillings only. The building, water-wheels, gearing, and machinery were of the latest and best construction at that time, and fully equal to those of any mill in Lowell.

"The machinery in the mill was as follows : Two conical willows. Two pickers or lappers, with 2 beaters each. Twenty-eight breaker-cards, with main cylinders 37 inches in diameter and 37 inches wide, with a leader-in $6\frac{3}{4}$ inches in diameter, and 12 top flats ;

15

draught, 32. Two lap-winders for making a lap for the finisher cards, from 32 breaker card slivers. Twenty-eight finisher cards, with main cylinders same size as the breaker cards, and with 14 top-flats; draught, 31.27. First set of drawing-frames had 16 deliveries, and doubled 3 into 1; draught, 4.12. Second set of drawing-frames had 24 deliveries, and doubled 4 into 1; draught, 3.76. Third set of drawing-frames had 24 deliveries, and doubled 4 into 1; draught, 3.17. Six speeders of 18 spindles each, having bobbins with heads $6\frac{1}{2}$ inches in diameter and 8 inches long between the heads; draught, 5.7; twist, .71 per inch. Ten fine speeders or stretchers, of 36 spindles each, having bobbins with heads 5 inches in diameter and 7 inches long between the heads; draught, 6.13; twist, 1.2 per inch. The coarse or speeder roving was doubled on the stretchers. Twenty-eight flier and dead-spindle throstles, of 128 spindles each, for making warp; 800 yards of yarn put on a bobbin. Twenty flier and dead-spindle throstles, of 128 spindles each, for making filling; 400 yards of yarn put on a quill; draught on all the throstles, 9.94. Three thousand five hundred and eighty-four warp and 2,560 filling spindles, making 6,144 spindles in the mill. Eight 'cradle' warpers, for putting 250 threads and 5,000 yards in length on a section beam. Twelve dressing-frames, carrying 8 section beams each, having 2 fans, and drying the sized yarn with the air of the room, or that coming from the furnace which warmed the mill. Ten cuts of 32 yards each, when wove, were put on a loom beam. One hundred and seventy-six looms for weaving drillings 30 inches wide.

"The comparison is made for four weeks (24 days) in May, 1838, and for the same length of time in May, 1876.

"The statistics for 1876 were very kindly given by Mr. A. G. Cumnock, the present agent of the Boott Cotton Mills.

"Since 1861 all the mills owned by the Boott Cotton Mills have been renovated and enlarged (one new mill added), supplied with additional motive power, new shafting, and an entirely new suit of machinery of the latest construction, arranged for the greatest economy in operating.

"The number of spindles in all the mills when they were started, in 1836-'37, was about 28,000. The present number of spindles is about 113,000, and of looms 2,550, while the capital stock is the same as in 1836, viz., $1,200,000.

"All these renewals and additions have been paid for from the earnings, and the mills are believed to be equal to any in New England for economy in working and in the quality of the cloth manufactured.

BOOTT COTTON MILL No. 1.

	MAY, 1838.	MAY, 1876.
Organization of Cloth (Drillings), viz.:		
Number of the yarn (average)	13.64	13.93
Threads in the warp	2,000	2,196
Picks of filling per inch	50	50
Weight in yards per pound	2.91	2.85
Hours of labor per week	76¼	60
Pounds of cloth made in 306 hours	71,686	....
Pounds of cloth made in 240 hours *	....	71,882
Number of looms used	176	194
Yards woven on a loom in 60 hours	245½	264
Number of spindles run	6,144	6,965
Pounds spun per spindle in 60 hours	2.292	2.58
Number of Operatives, viz.:		
In card room (including picking)—		
Males	14.3	9.33
Females	33	11
In spinning room—		
Males	4.18	2.5
Females (including spoolers)	55	25
In dressing room—		
Males	2	1.5
Females (including warper tenders)	29	4
In weaving room—		
Males	3	2.5
Females	86	34
Total Males	23.48	15.83
Total Females	203	74
Total operatives	226.48	89.83
Pounds of cloth produced by each operative in one hour	1.012	3.333
Cost of labor per pound—	Cents.	Cents.
For picking, carding, and roving	1.0291	.6674
For spinning (including spooling)	1.1168	.7446
For warping and dressing	.7105	.1786
For weaving	1.9371	1.2627
Total cost of labor per pound	4.7935	2.8533
Average of Wages paid (Board included, Overseeing excepted).		
In picking and carding rooms—		
For males, per day	$0.76¼	$1.22¼
For females, per week	3.02½	3.98 7/10

* On account of the changes made in the machinery and its position, the pounds of cloth given as made in May, 1876, are 193 more than were made in May, 1838.

BOOTT COTTON MILL No. 1.—(*Continued.*)

	MAY, 1838.	MAY, 1876.
Average of Wages paid (Board included, Overseeing excepted).		
In spinning room—		
For males, per day.........................	$0.66¾	$1.00
For females, per week (including spoolers)..........	2.93½	4.27½
In dressing room—		
For males, per day.........................	.66¾	1.25
For females, per week (including warpers and web-drawers.........................	3.62	5.40
In weaving room—		
For males, per day.........................	.72½	1.00
For females, per week.......................	3.39½	5.88
Price of board in corporation houses—		
For males...............................	1.75	3.25
For females..............................	1.37½	2.10

NOTE.—In addition to the $2.10 per week paid by females for board in 1876, the company pays 30 cents per week to the boarding-house keeper, making $2.40 in all.

"The mill of the Jackson Company was put in operation in 1832, but, as the accounts for the first year were not kept in detail, the *six* months ending June 1, 1838, and the *six* months ending April 30, 1876, are taken for comparison.

JACKSON COMPANY.

	1838.	1876.
Number of spindles.........................	12,000	23,888
Number of looms...........................	400	786
Average number of yarn.....................	13.25	13.25
Weight of cloth in yards per pound............	2.95	2.932
Cost of labor per pound.....................	4.805 cts.	3.50 cts.
General expenses per pound..................	2.137 cts.	2.605 cts.
Cost of cotton per pound at the mill...........	12.73 cts.	14.132 cts.
Percentage of waste, net....................	12.91	12.11
Total cost of cloth per pound.................	21.99 cts.	22.289 cts.
Total cost of cloth per yard..................	6.64 cts.	7.601 cts.
Total cost of print cloth per yard.............	5.726 cts.	None made.
Yards of cloth made in 6 months..............	2,832,575	4,737,681
Pounds of cloth made in 6 months.............	960,195	1,615,791
Average price per yard received for sales......	8.50 cts.	8.549 cts.
Profit per yard, net.........................	1.86 cts.	.948 cts.
Pounds spun per spindle in 64¼ hours.........	2.71	2.635
Pounds woven per loom in 64¼ hours..........	81.51	80.09
Yards woven per loom per day of 11 hours.....	41.03	39.14

JACKSON COMPANY.—(*Continued.*)

	1838.	1876.
Six months is equal to 154 days, or 25⅔ weeks of 6 days each.—Hours of labor per week, say..................	74	64½
Operatives employed—		
Males..	514.62 } in all.	62.64
Females..		352.40
Hours of labor for 6 months = 154 days = 25⅔ weeks of 6 days each...................................	1,898.84	435.04 1,655.07
Cloth in pounds produced in one hour by each operative...	.9852	2.275

"When the mill began work, it had 10,240 spindles and 360 looms. No record of the machinery appears to have been kept until June 1, 1841, when the number of spindles is given as 12,500. I assume that in 1838 there were 12,000 spindles and 400 looms. For a few years the company made a small amount of print cloths, but they were discontinued, and the whole product of the mill has been 'Indian Head' standard sheetings, so favorably known over the whole country.

"The statistics were kindly furnished by Frederic Amory, Esq., of Boston, the treasurer of the Jackson Company.

"The cost of labor, the number of operatives, and the average of wages paid in each department could not be ascertained for the six months in 1838, and of course are omitted for the six months in 1876.

"As the reduction in the cost of labor and the number of operatives employed does not appear to be as great at the Jackson Company as at the Boott Cotton Mills, it is but fair to remark that the Jackson Company continue to use the flier and dead-spindle throstle for spinning (most of it the same as when the mill was started), and have not until recently substituted 'slashers' for the old-style 'dresser' in sizing the warp yarn.

"My purpose is to show what progress has been made as regards the labor-cost of making drillings and standard sheetings from 1838 to the present time. Although the comparison is made on No. 14 yarns, yet I think an equally favorable result would be found on finer numbers, say as fine as No. 30.

"As regards prices for labor, we know that is always affected by the demand and supply of labor. Wages are now greater than they were thirty-eight years ago, but not as large as within the past few years.

"The wages as given at the Boott Cotton Mills, after deducting the prices paid for board by males and females, show at the present time an increase in the wages of males (overseers are not included) of sixteen and one half cents per day, and of females of eighty-nine cents per week, more than the net wages received by them in 1838, or an increase of 40 per cent. for males and 47 per cent. for females.

"At the Boott Cotton Mills the labor-cost on drillings is 1.94 cent per pound less in 1876 than in 1838, and is distributed as follows:

```
Less in card room........................................ .3617 cent.
"    in spinning room.................................... .3722  "
"    in dressing room.................................... .5319  "
"    in weaving room..................................... .6744  "
                                                         ───────
                                                         1.9402  "
```

"But a greater difference appears in the amount of cloth produced by each operative. This is shown by the fact that at the Boott Cotton Mills one hour's work by each person gives 3.33 pounds of cloth in 1876 and but 1.012 pound in 1838.

"At the Jackson Company, in 1876, one hour's work gave per hand 2.275 pounds of cloth, and in 1838 but .9852 pound. I presume the records of other mills would show equally well.

"How has this improvement been obtained? I will specify a few of the changes that have taken place since 1838, as they appear from my observation:

"First. Larger mills, with better opportunity for arranging machinery to economize labor. The size of cotton mills, as established at Lowell forty years ago, was 6,144 spindles for No. 14s and about 8,500 spindles for No. 30s, and the machinery for weaving the yarn into cloth. At this time a 20,000 spindle mill is a moderate size, and we have them of 50,000 spindles, or even more.

"Second. Improvements in the construction and workmanship of machinery and many important inventions and attachments to save labor and perfect work. I will note but some of the principal ones: The Wellman Top Card Stripper, the use of lap-heads (so called) where double carding is practiced, eveners on railway-heads, the stop-motion on drawing-frames, great improvements on mules, the introduction of the ring and traveler spinning-frames, also of the 'slasher' for sizing yarn, and the filling stop-motion on the loom. These are but few of many improvements familiar to all of us.

"Third. The number of looms a weaver is now able to tend has more than doubled. In 1838 two looms to a weaver was the rule, though there were cases of three or more being tended by one person. Now the practice is for four to six and even eight looms to be run by one weaver. At the Boott Mills 34 weavers tend 194 looms ; and, if two of the 34 are "room girls," then 194 by 32 would give a trifle more than six looms to a weaver.

"Fourth. The reduction of at least one half of the piecings in the progress of the cotton from the bale to the cloth. We now make longer laps and use larger cans for the drawing-sliver ; by improvements on fly-frames and on speeders, we double at least the length of roving laid on a bobbin, and thus enable a spinner to tend more spindles. We double the length of yarn wound on a quill or bobbin ; we wind three times as much weight of yarn on a 'section' or 'slasher' beam, and we double at least the number of cuts or pieces on the warp beam for the loom.

"These are the principal changes that occur to me as having operated to increase so largely the amount of cloth made by each person employed. They are the result of the experience and persistent labor of many years, have been of slow growth, and obtained by a little here and a little there.

"I might give another reason for the progress made, especially within the last fifteen or twenty years. We are more sensible of the advantage of keeping well informed and 'posted' in whatever relates to economy in cost of production, and for that object there is now a greater interchange of information of what is being done in our mills than was formerly the custom.

"This association is an important aid in that direction, and now, in the eleventh year of a flourishing and I hope permanent existence, is carrying out the purpose set forth in the preamble to its constitution, viz.: 'Promoting a more intimate acquaintance with each other, and collecting and imparting information as to the best methods of manufacturing cotton.'"

APPENDIX B.

LETTER FROM AZA ARNOLD TO THE CHAIRMAN OF THE COMMITTEE ON
PATENTS, UNITED STATES HOUSE OF REPRESENTATIVES.

"WASHINGTON, *September* 6, 1861.

"HON. ELISHA DYER, *Chairman, etc.:* We are informed that the differential speeder is claimed by Mr. Appleton as a Waltham invention. But the author of Waltham inventions made no such claim. No improvement on cotton machinery appears to have been made at Waltham, up to 1826, except by Paul Moody; he was chief mechanician of Waltham, and claimed to be the inventor of the Waltham speeder; he claimed eight improvements on the machine, but they proved not to be new. Jonathan Fisk also built the same kind (Waltham speeder) at Medway, and took five patents on the machine. William Hines, of Coventry, R. I., had made improvements on the speeder and patented before them. And it is remarkable that the parts claimed by Moody are the identical parts which are superseded by my compound motion, and were never used in a differential speeder. I shall refer to the case of Moody *vs.* Fisk in a future page, to show that Moody's claim proves the Waltham speeder to be essentially different. Paul Moody took charge of the Lowell establishments, and Jonathan Fisk took charge of the Dover factory. And I shall show that neither Moody nor Fisk knew any method of compounding two different motions, and producing their differential, for four years after I had the machine in operation. I was well acquainted with Moody, saw his machines, and considered that he improved the speeder by adopting the long flier; but the long flier was invented by Asa Gilson, at Dorchester. I have used both Hine's and Fisk's speeders, and well remember the difference.

"If I exhibit a little egotism in this reminiscence, you will excuse it when you consider the local prejudice that was exercised against my machine as a Rhode Island invention. I invented the differential speeder, and put it in operation in 1822, at South Kingstown, and it was soon in operation at Coventry, Scituate, Pomfret, and a dozen other places, but for three or four years it was discountenanced at Waltham, and Lowell, the Waltham speeder being exclusively used in both places until I had constructed and put in operation the Great Falls factory, at Somersworth, N. H., which actually produced 30 per cent. more goods per week than the Waltham or Lowell factories had produced, of equal quality. This brought down the directors of the Lowell factories to our place at Somersworth, to inquire into the cause of so great a difference. It brought also Mr. Moody, their engineer, and Mr. George Brownell, the foreman of the Lowell machine shop ; they also sent the celebrated mathematician, Warren Colburn, to see if our calculation was correct. I had the pleasure of exhibiting and explaining all the minutiæ of the Rhode Island invention a third time, and the result was that Mr. Colburn told Moody that it was mathematically correct, and that it was the only plan that he had heard of by which the machine could be made adjustable to all sizes of ropings. We notice the case Moody *vs.* Fisk (2 Mason Rep., 112), tried at Boston, October term, 1820. In the defense, it was proved that the improvements claimed by Moody were not new, neither were they invented at Waltham. William Hines, of Coventry, had made improvements on the speeder, and patented in February, 1819, previous to Moody's date. Moody's patent was vacated for want of novelty. The object of referring to it is to show that Moody's claim proves the Waltham speeder to be a different machine from the differential speeder. In summing up his claims, he says : "*First*, I claim the position of the rolls. *Second*, the two upper cones. *Third*, the method of moving the belt on the two lower cones, and that of communicating motion from the lower cones to the spindles, and all the mechanism and method of communicating motion, from the upper driven cone to the arbors or axles of the endless screws, and perpendicular racks or screws that raise, and the spindle rail. *Fifth*, I claim the method and machinery by which the said motion communicated to the spindle rail is changed from an ascending to a descending motion, and the manner of connecting the same with the wagon carriage. *Sixth*, the wagon and the wagon carriage, gallows frame, catch wheel, the cycloid cam, slide lever and pulley shaft, which raises the belt on the upper cones, and all the similar parts that raise the belt on the lower cones (except the cycloid, or cycloid cam), with all the parts, movements, and mechanism connected with the same. *Seventh*, the flier tubes, and method of applying and using them. *Eighth*, the rotary motion of the cams, and the intermediate gear work. And further, I claim that these my inventions are applicable, not only to this machine which is adapted to one size of roping, but may be proportioned and applied to the making of any other kind of roping." So, by his own showing, the Waltham speeder makes but one size of roping. It is proper to remark that my compounding wheels supersede all the second pair of cones, cycloid cams, the cycloid racks, the second cone belt, and the method of moving the belt, which required to be brought up by ratchets and catches, with teeth of different lengths, graduated to suit one size of roping, and which could not be used to make a different grade of roping, finer or coarser, but require another set of parts graduated differently to suit any other size of roping, and this proportioning and adjusting of the machinery was required at each change from fine to coarse, or from coarse to fine. The object of my inventing the differential speeder was to do away with the intricate construction, and to simplify and extend the use of the machine, so that one set of gears can be adjusted to each and every size of roping by merely changing the pinions. When Mr. Moody came to me for an explanation of my invention, we had a free and full discussion of its parts and properties. I remarked to him that the exact difference between the retarding motion and a certain uniform motion would be always right for the accelerating motion. He seemed not to recognize the fact, and spoke doubtfully of it; I then remarked that the same cause that required the graduating of one, required the graduation of the other, for both depend on the diameter of the roping. Therefore, I take the advantage of using this differential for the accelerated motion, rather than to use another pair of cones and belt fixtures; but I have another more important advantage by so doing, that is, whenever it is required to alter one graduation, the other always keeps right along with it; whatever may be the rate of change required, these motions are always reciprocal to each other. Therefore, I use a rack with equal teeth for moving the belt, and move it by a pinion of any requisite number of teeth, so as to adapt the same machine to any size of roping by merely changing the pinions.

"Up to this time, the differential speeder had not been seen at Waltham or Lowell, neither had the authors of Waltham inventions taken the pains to investigate its merits. But after this, I had a cordial and good understanding with both Moody and Fisk. I have subsequently been informed by Mr. George Brownell that, soon after this interview, they commenced making my kind of gears at Lowell, and not only built my kind of speeders, but also took up their Waltham speeders, and geared them over, and converted them into differential speeders, by putting in my compound motion. This is a historical fact of some significance; George Brownell, I think, is still living at Lowell, and will confirm these remarks; James Dennis, Gideon C. Smith, and Daniel Osborn, who were with us at Somersworth, may, perhaps, recollect some of the circumstances. While on the subject we may remark further that the speeder (fly frame) had been used in England, but the compound motion or differential had never been applied to an English machine, until Charles Richmond carried to England a model of my wheels (unbeknown to me). He was there in 1824–5, when Mr. Houldsworth took up the subject of improving the fly frame. Dr. Ure informs us that Houldsworth applied the differential system and patented it in 1826; that is, three years after the date of my patent. It was not requisite for him to claim it as his original invention. I have been informed, through a former partner of Charles Richmond, that the model which he carried to England was made in Taunton, and was sold in England, and had since been patented there. We said that J. Fisk did not understand producing and using the differential motion until three years after we had the machine in operation. It happened that J. Cowing, in describing my speeder, told Fisk that it had but one pair of cones, and one cone belt. Fisk remarked, then it could not work. Cowing replied, "but it appears to work right well, and makes more roping than the Waltham speeder." Mr. Fisk then entered into argument, saying, "It is impossible to produce both graduations by one pair of cones and one belt, because, while one is a retarding motion to vibrate the spindle rail, the other requires to be an accelerated motion for the winding up." So it was evident that he did not understand it, or he would not have made this assertion. If my differential speeder had ever been supposed to have been a Waltham invention, we should have heard of it during my three years' contest with six corporations of Lowell, yet not a word of any such claim was offered, but, on the contrary, they tacitly acknowledged my right to the invention; and after having the law repealed, thereby defeating my first claims up to that time, they then gave me $3,500 for the right to use the same for the last year of the term of my patent. And this they did after searching all the evidence that could be found against my claim. Mr. Lyman, of Boston, who acted as their agent, who paid me the money and received the license for them, told me they found no evidence against it. Few readers will take the trouble to understand the specific difference between two complex machines; but when one mode of operation enables the manufacturer to produce twenty per cent. more goods, with the same cost of labor than has before been done, it becomes of national importance. Dr. Ure well remarks that, since the differential system has been adopted, manufacturers have been able to produce a better article at a less cost, and have thereby increased the trade.

"I am, dear Sir, most respectfully,
"Your friend and servant,
"AZA ARNOLD."

DISTINGUISHING EXCELLENCES
OF
THE AMERICAN CYCLOPÆDIA.

I. ACCURACY AND FRESHNESS OF INFORMATION.—The value of a work of this kind is exactly proportioned to its correctness; and to insure that, as well as the latest information, no expense or literary labor has been considered too great.

II. IMPARTIALITY.—The work has been pronounced by distinguished men and leading reviews in all parts of the Union strictly fair and national. Eschewing all expressions of opinion on controverted points of science, philosophy, religion, and politics, it has aimed at an accurate representation of facts and institutions, of the results of physical research, of the prominent events in the history of the world, of the most significant productions of literature and art, and of the celebrated individuals whose names have become associated with the conspicuous phenomena of their age, doing justice to all men, all creeds, all sections.

III. COMPLETENESS.—It treats of every subject, in a terse and condensed style, but fully and exhaustively.

IV. ITS AMERICAN CHARACTER.—THE AMERICAN CYCLOPÆDIA is found especially to meet the intellectual wants of the American people. It is not, therefore, modeled after any European works; but, while it embraces all their excellences, it has added to them an unmistakable American character, and is the production mainly of American mind.

V. ITS PRACTICAL BEARING.—The day of philosophical abstraction and speculation has passed away. In this age of action, the work has been made thoroughly practical.

VI. ITS STYLE.—The cold, formal, and repulsive style of older works of this kind has given place to a style sparkling and emphatically readable—a style that interests and pleases as well as instructs. Many of the writers hold the foremost rank in general literature, and their articles have been characterized by the best critics as models of elegance, force, and beauty.

VII. CONVENIENCE OF FORM.—No ponderous work, with small type that strains the eyes and wearies the brain, is here presented. The volumes are just the right size to handle conveniently, the paper is thick and white, the type large, the binding good and durable.

VIII. CHEAPNESS.—The work has been regarded as a miracle of cheapness. In order to enlarge its sphere of usefulness, and make it, in a true sense, a book for the people, it is offered to the public at the lowest possible price.

To sum up the above and other advantages of this work, it claims that it surpasses all other works in the fullness and ability of the articles relating to the United States; that no other book contains so many reliable biographies of the leading men of this and other nations; that the best minds of the country have enriched its pages with the latest data, and the most recent discoveries in manufactures, mechanics, and general science; that it is a library in itself; that it is well printed, and in convenient form; that it is reliable, impartial, complete, thoroughly American, deeply interesting and instructive, and cheap.

D. APPLETON & CO., Publishers, 1, 3, & 5 Bond St., New York.

THE
HISTORICAL REFERENCE-BOOK,

COMPRISING:

A Chronological Table of Universal History, a Chronological Dictionary of Universal History, a Biographical Dictionary.

WITH GEOGRAPHICAL NOTES.

FOR THE USE OF STUDENTS, TEACHERS, AND READERS.

By LOUIS HEILPRIN.

New edition. Crown 8vo. Half leather, $3.00.

"A second revised edition of Mr. Louis Heilprin's 'Historical Reference-Book' has just appeared, marking the well-earned success of this admirable work—a dictionary of dates, a dictionary of events (with a special gazetteer for the places mentioned), and a concise biographical dictionary, all in one, and all in the highest degree trustworthy. Mr. Heilprin's revision is as thorough as his original work. Any one can test it by running over the list of persons deceased since this manual first appeared. Corrections, too, have been made, as we can testify in one instance at least."—*New York Evening Post.*

"One of the most complete, compact, and valuable works of reference yet produced."
—*Troy Daily Times.*

"Unequaled in its field."—*Boston Courier.*

"A small library in itself."—*Chicago Dial.*

"An invaluable book of reference, useful alike to the student and the general reader. The arrangement could scarcely be better or more convenient."—*New York Herald.*

"The conspectus of the world's history presented in the first part of the book is as full as the wisest terseness could put within the space."—*Philadelphia American.*

"We miss hardly anything that we should consider desirable, and we have not been able to detect a single mistake or misprint."—*New York Nation.*

"So far as we have tested the accuracy of the present work we have found it without flaw."—*Christian Union.*

"The conspicuous merits of the work are condensation and accuracy. These points alone should suffice to give the 'Historical Reference Book' a place in every public and private library."—*Boston Beacon.*

"The method of the tabulation is admirable for ready reference."—*New York Home Journal.*

"This cyclopædia of condensed knowledge is a work that will speedily become a necessity to the general reader, as well as to the student."—*Detroit Free Press.*

"For clearness, correctness, and the readiness with which the reader can find the information of which he is in search, the volume is far in advance of any work of its kind with which we are acquainted."—*Boston Saturday Evening Gazette.*

"The latest dates have been given. *The geographical notes which accompany the historical incidents are a novel addition, and exceedingly helpful.* The size also commends it, making it convenient for constant reference, while the three divisions and careful elimination of minor and uninteresting incidents make it much easier to find dates and events about which accuracy is necessary. Sir William Hamilton avers that too retentive a memory tends to hinder the development of the judgment by presenting too much for decision. A work like this is thus better than memory. It is a 'mental larder' which needs no care, and whose contents are ever available."—*New York University Quarterly.*

New York: D. APPLETON & CO., Publishers, 1, 3, & 5 Bond Street.

The history of a people is largely composed of the biographies of its great men, and the reader or writer of history needs a biographical dictionary constantly at his elbow.

APPLETONS' CYCLOPÆDIA OF AMERICAN BIOGRAPHY

Contains a biographical sketch of every person eminent in American civil and military history, in law and in politics, in divinity, literature and art, in science, and in invention, including distinguished persons born abroad that are related to our national history, and embraces all the countries of North and South America.

From the Hon. GEORGE BANCROFT.

"The most complete work that exists on the subject."

From the Hon. JAMES RUSSELL LOWELL.

"Surprisingly well done. . . . To any interested in American history or literature, the work will be indispensable."

From NOAH PORTER, D. D., LL. D., *ex-President of Yale College.*

"It is with great pleasure that I certify to the excellence of 'Appletons' Cyclopædia of American Biography.'"

From the Hon. M. R. WAITE, *Chief-Justice of the United States.*

"I have looked it over with considerable care, and find nothing to say except in praise."

This great national work will be completed in six volumes royal octavo, of nearly 850 pages each.

☞ Full descriptive prospectus with specimen pages will be sent to any address on application.

Sold only by subscription. Agents wanted for districts not yet assigned.

New York: D. APPLETON & CO., 1, 3, & 5 Bond Street.

SCIENCE AND TRAVEL.

TWO NOTABLE VOLUMES.

A NATURALIST'S VOYAGE ROUND THE WORLD.

JOURNAL OF RESEARCHES INTO THE NATURAL HISTORY AND GEOLOGY OF THE COUNTRIES VISITED DURING THE VOYAGE ROUND THE WORLD OF H. M. S. 'BEAGLE.' By CHARLES DARWIN. *New illustrated edition.* With Maps and 100 Views of the places visited and described, chiefly from sketches taken on the spot, by ROBERT TAYLOR PRITCHETT. One vol., 8vo. Cloth, $5.00.

The object of this edition is to aid the author's descriptions by actual representations of the most interesting places and objects of natural history referred to in them. This has been effected by securing the services of an artist who has visited the countries which Darwin describes. Most of the views are from sketches made on the spot by Mr. Pritchett (well known by his connection with the voyages of the *Sunbeam* and *Wanderer*), with Mr. Darwin's book by his side. Some few of the others are taken from engravings which Mr. Darwin had himself selected for their interest as illustrating his voyage, and which have been kindly lent by his son.

"One of the most interesting narratives of voyaging that has fallen to our lot to take up, and one which must always occupy a distinguished place in the history of scientific navigation."—*Quarterly Review.*

THE ICE AGE IN NORTH AMERICA,

AND ITS BEARINGS UPON THE ANTIQUITY OF MAN. By G. FREDERICK WRIGHT, D. D., LL. D., F. G. S. A., Professor in Oberlin Theological Seminary; Assistant on the United States Geological Survey. With an Appendix on THE PROBABLE CAUSE OF GLACIATION, by WARREN UPHAM, F. G. S. A., Assistant on the Geological Surveys of New Hampshire, Minnesota, and the United States. With 147 Maps and Illustrations. One vol., 8vo, 640 pages. Second edition. Cloth, $5.00.

The numerous maps accompanying the text have been compiled from the latest data. The illustrations are more ample than have ever before been applied to the subject, being mostly reproductions of photographs taken by various members of the United States Geological Survey in the course of the past ten years, many of them by the author himself.

"The special study which Dr. Wright has made of this era, the peculiar facilities which he has enjoyed as an assistant in the United States Geological Survey, and the habit of clear exposition which as a professor in a seminary like that of Oberlin naturally acquires, have enabled him to produce a work worthy of the importance and interest of his subject. It is not always, or indeed often, that a work of pure science can be made both instructive and attractive to readers not familiar with the principles of the science involved In this instance, however, the subject naturally lends itself to what may be styled a popular treatment; and the author has aided his explanations by a profusion of maps and pictures, the latter mostly photographic, which render his descriptions and the consequent inferences plain to any reader of ordinary intelligence."—*The Critic.*

D. APPLETON & CO., Publishers,
1, 3, & 5 Bond Street, New York.

SOUTHEASTERN MASSACHUSETTS UNIVERSITY
TS1583.W37
Manual of power for machines, shafts, an

3 2922 00084 412 3

DATE DUE

HIGHSMITH #45230

www.ingramcontent.com/pod-product-compliance
Lightning Source LLC
Chambersburg PA
CBHW020800230426
43666CB00007B/787